THE MAN WHO TOUCHED HIS OWN HEART

Also by Rob Dunn

The Wild Life of Our Bodies
Every Living Thing

THE MAN WHO TOUCHED HIS OWN HEART

*True Tales of Science,
Surgery, and Mystery*

ROB DUNN

LITTLE, BROWN AND COMPANY
New York Boston London

Little, Brown and Company
Hachette Book Group
1290 Avenue of the Americas, New York, NY 10104
littlebrown.com

First Edition: February 2015

Little, Brown and Company is a division of Hachette Book Group, Inc.
The Little, Brown name and logo are trademarks of Hachette Book Group, Inc.

The publisher is not responsible for websites (or their content) that are not owned by the publisher.

The Hachette Speakers Bureau provides a wide range of authors for speaking events. To find out more, go to hachettespeakersbureau.com or call (866) 376-6591.

Library of Congress Cataloging-in-Publication Data

Dunn, Rob R.
    The man who touched his own heart : true tales of science, surgery, and mystery / Robert Dunn.
        pages cm
    ISBN 978-0-316-22579-3 (hardcover)
    1. Heart—History.   2. Cardiology—History.   3. Heart—Diseases—History.   4. Heart—Surgery—History.   I. Title.
    QP111.4.D86 2015
    612.1'7—dc23   2014019901

RRD-C

10 9 8 7 6 5 4 3 2 1

Printed in the United States of America

To Monica, Lula, and August,
for whom my heart beats

I counsel you not to cumber yourself with words unless you are speaking to the blind.... How in words can you describe this heart without filling a whole book?

— LEONARDO DA VINCI

Contents

CONTENTS

The Human Heart

Today, one in three adults in the world will die of a disease of the cardiovascular system: a stroke, a heart attack, or another disorder of the heart, blood, arteries, or veins. In children, the most common congenital diseases are those of the heart. In the coming years as the Western world ages and the rest of the world begins to better escape (one can only hope) pathogens and parasites, diseases of the heart will be even more common. Our hearts are our weakness. They are also our strength. This is their story.

THE MAN WHO TOUCHED HIS OWN HEART

Introduction

Heart diseases are so common that it was almost inevitable that someone I knew well would suffer from a problem heart while I was writing this book. I just didn't think it would be my mother. On January 4, 2013, my mother, who lives in Wasilla, Alaska, went to her doctor. Her doctor took her blood pressure, ran an electrocardiogram, and then took immediate action. My mother's heartbeat was irregular (arrhythmia) and her heart rate was extraordinarily high (tachycardia)—184, the sort of heart rate more often associated with a small bird than a human. Her blood pressure was up too. There was no telling how long she had lived with these conditions. Months? Years? Immediately, my mother was put on a cocktail of drugs to lower her heart rate and blood pressure and to sort out the wriggling arrhythmia. Both of the problems from which my mother was suffering at this point in the story are common. Arrhythmia affects as many as one out of three adults over sixty, tachycardia fewer, but still hundreds of thousands. This very commonness was, to a son many miles away, reassuring.

What was not reassuring was that the cocktail of drugs that were given to my mother did not seem to fix her problems, at least initially. Slowly, her heart rate began to come down. This seemed like progress, but her arrhythmia did not go away; it seemed to get worse. Inside her body, her heart flopped and then flopped again.

On January 15, my mother was scheduled for a more extreme intervention. Doctors were going to shock her heart. They would stop it in the way that lightning can, electrically. The hope was

3

that, when her heart's rhythm resumed, it would be normal. Such a procedure is the medical equivalent of kicking the TV when it flickers in the hopes that whatever is loose will reconnect itself. It works about half the time.

My mother was terrified. The doctors seemed to have been told by someone to avoid using the word *shock* in describing the treatment. My mother asked, "So you will shock me?" The doctors replied, "No, no, it is not a shock. Do not worry." No one, however, told the technician about the party line, and so, in the moments before the procedure, he asked my mother, "Are they going to shock you? They shock people all day every day in here." Lightning is lightning, whatever you call it. They shocked my mother's heart. It stopped, restarted, and then resumed the fumbling beat it had had before the procedure, still arrhythmic.

My mother went home, her heart slowed but still skipping beats, sending blood irregularly through her body. She was worn out. Maybe she had been worn out for a while and had not realized it. She slept for twelve, and then more, hours a day. It could have been, it seemed, the effect of not getting enough blood through the body. It could have been other things. It was, it would turn out, other things.

On February 5, after about a month of feeling terrible (and probably a number of months before that of feeling poorly but not knowing why), my mother went to the doctor again. When she did, they took her blood pressure. They measured her heart rate. They ran an electrocardiogram, an EKG. Once again, the doctors, this time a different set, were alarmed. My mother was admitted to the ICU.

My mother had, unwittingly, been given too much of one of her drugs by her first heart doctor. The first doctor had prescribed a version of digitalis. Digitalis lowers heart rate, but its effects are highly dependent on its dosage. Too little, and it has no effect; a hair too much, and it can prove dangerous, deadly even. My mother had been given too much. The first symptom was yellowed vision (everything seen as if through amber-colored glasses). The second, although she didn't know it was a symptom, was sleepiness, an intense sleepiness that made her sleep at first ten and then twelve and then ultimately sixteen or so hours a day. The third symptom was a lack of appetite; my mother, it turned out, was not eating much at all and had been losing weight quickly. Then there were cognitive problems that reached their peak on the day my father took her back to the doctor. She could barely form words and even when she could, she could not seem to put them in the right order. Digitalis, a drug that should have made her better, had become her poison.

In the ICU, doctors put my mother on four IVs. They watched her constantly. They did test after test. Nothing seemed to make her better. It took a while for them to realize that all of her symptoms, apart from the heart-rate problems, were due to the digitalis, and even some of the heart-rate problems seemed to be related to the digitalis. Whereas her heart had once beaten too fast, it now beat too slowly, far too slowly, akin to the rate of an elephant: *thump, da, thump, da.* And the original arrhythmia was still there, but now worse.

Before the symptoms of poisoning emerged, the doctors planned to treat my mother's arrhythmia by ablation, a procedure in which the part of the heart responsible for the abnormal rhythm is destroyed. The hope is that this destruction of heart tissue will stop extraneous signals from causing the heart to misfire. Ablation is primitive medicine. It works, but we don't understand it well. In this, it is like shocking the heart.

But the primitiveness of the ablation would not matter to my mother for the simple reason that, after the digitalis poisoning, the condition of her arrhythmia was viewed as too extreme—too tenuous, too erratic—to be fixed by a burn to the heart. She would, the doctors agreed, need a pacemaker if she got better. Suddenly, the doctors were talking about *if* she got better.

She did get better, slowly. As the digitalis was washed out of her system, my mother's odds of getting a pacemaker went up and up. Her magnesium levels were getting closer and closer to normal (though they would stay high only if she continued to be on a drip). Her potassium went up too. Slowly. And maybe, it seemed, maybe her cognitive impairment was diminishing. Though maybe not. By February 10, five days after she entered the ICU, my mother was deemed well enough for a pacemaker.

Unfortunately, the cardiologists at her small hospital were not equipped to implant a pacemaker. She would have to move. But there weren't any beds at the nearest large hospital, in Anchorage, where pacemakers were regularly implanted. She waited, we waited, for hours, then days. I contemplated flying to Alaska to drive her to...well, there was nowhere else.

After several days, thanks to a change in someone else's story—an improvement, a death; who can say?—a bed opened up. She was driven to Anchorage. Once she was settled in, the doctors operated. Though it was hardly an operation. A small slit was made in the skin beneath her collarbone, and through it a catheter was inserted into her left subclavian vein. The catheter was threaded through the vein and into the right side of her heart, where it was followed by another piggybacking catheter, this one carrying an electrode that would be implanted directly on her heart, an electrode to which the pacemaker would send signals telling it to *beat, beat, beat.* Here, in the pacemaker, was unbelievable modernity. The tiny device contained a rhythm, a battery, an artificial piece of humanity to be inserted into her body without major surgery, without ever really opening her up. It was a device that would run her rhythm, if everything worked, for the rest of her days, with the only necessary tweak being the potential need for a new battery in five or ten years.

The pacemaker was implanted under the skin of my mom's chest. Then, later that day, she, unceremoniously and half mended, was sent home to be with her cats, her dog, and my father. Everyone hoped she would continue to improve.

My mother's story is simultaneously terrible, ridiculous, fraught, and modern. It is also, in many ways, typical. Tachycardia and arrhythmia are two of the dozens of heart problems common today; nearly all of us will encounter at least one of them at some point. Cardiovascular diseases, particularly strokes, are the most common cause of death in the United States and in most developed countries around the world. And even when there are ways to treat these deadly problems before they kill us, such as with my mother's arrhythmia, they do not disappear entirely. They remain chronic weaknesses, the fragility just beneath the skin's surface.

Here I tell the story of the heart. I confront the question of why our hearts—ones like my mother's, but also yours—break so often, more than any other part of our bodies. Our heartbreak is an ancient story, one that begins hundreds of millions of years ago when our ancestors were just single cells, but the story of the science of the heart is more recent. It begins just six thousand years ago. As for the mending of our hearts, that does not begin until the end of the nineteenth century, when the very first nick to a living heart was made with the aim of repair, a nick that led to all the subsequent ones, a nick with a knife nearly identical to the one that cut into the skin of my mother. Then there are the mysteries of the heart, mysteries we have only begun to unravel, mysteries at the center of who and what we are.

In the 1400s, it was often said that the story of each lived life was written on the inside walls of the heart by a scribbling and obsessive God. When the heart was finally opened and examined in detail later in the same century, no such notes were discovered. Still, each mended heart bears the mark of a different kind of narration. Each mended heart beats out a conclusion to the struggles of the scientists, artists, surgeons, and writers who, with heroism, hubris, and insight, have done battle with the heart's mysteries for millennia. Each mended heart beats out a story of frailty but also of possibility.

As for the specific conclusion to my mother's story, it is, like many others', a patchwork and only partially resolved. It is a patchwork in that the mending of her heart depended on a mix of incredibly sophisticated tools as well as more ancient ones—the pacemaker and the stethoscope, 3-D scans and cauterization. But it is a patchwork that also reflects an approach in which the symptoms are tended to with very little attention to why things have gone wrong. Her story, in this way, is reflective of the broader story of the heart, a story in which every person's heart is influenced by ancient problems that have to do with the heart and the heart's evolution meeting up with modern circumstances.

My mother's story is only partially resolved in that she continues to recover. When she left the hospital, she was still very sick, but her heart rate was suddenly just right, eighty beats per minute, beats set by a tiny pulse of electricity from the new pacemaker, her own personal lightning. She was still weak. Her words still failed her. She was still dizzy. Her potassium was still far too low, as was her magnesium. Slowly, though, things seemed to get better. After a week, she could converse normally, more or less. By two weeks, she felt as good as she had in a month. By three weeks, she felt better than she had before the whole ordeal started. Now, she says she feels better than she has in years. Her improvement is due to the understanding we have gained of our most central organ, an understanding that remains humble and yet has advanced enough that my mother is back up and walking around, advanced enough that many people with similar problems are up and walking around— impelled by the collective lightning of pacemakers, discovery, mechanical parts, and much, much more.

I could begin to tell the heart's stories at any moment in the past, either four billion years ago or four seconds ago, but the odyssey of the heart pivots on a moment one hot day in 1893 when a man in a poor hospital in a rough part of Chicago decided, for what seemed to be the very first time in history, to cut into a heart in order to heal it. It took nearly six thousand years for scientists and practitioners to understand enough for him to lift that blade, cutting loose a bold era of discovery that continues on into this century, one that depends on an understanding of biology, evolution, art, plumbing, nuclear physics, and nearly everything else. In the end, more so than for any other part of the body, understanding the heart has required every tool that humanity has developed, and even so, the meat in the middle of you, pounding right now, is only partially understood.

1

The Bar Fight That Precipitated the Dawn of Heart Surgery

Any surgeon who would attempt an operation of the heart should lose the respect of his colleagues.

— T. H. BILLROTH, GERMAN SURGEON

It was July of 1893, and the city of Chicago was melting. It was the summer of the World's Fair, when inventions from around the world began to transform America. By fall, the first hamburger would arrive in Chicago, as would the first machinery for making chocolate commercially and the first tinny version of Alexander Graham Bell's phone. It was also the summer in which Daniel Hale Williams (1856–1931), a young doctor from the rough side of town, would make the biggest decision of his life.[1]

Williams was born of African American–Scots–Irish–Shawnee parents, but he was viewed by the society in which he lived, the society of Hollidaysburg, Pennsylvania, as African American. Williams's father died when he was young, leaving his mother to care for him alone. She was sufficiently overwhelmed that she sent Daniel to be an apprentice to a shoemaker in Baltimore when he was just eleven. That might have been the end of the story, except that young Williams decided to go to Wisconsin, where he began working in a barbershop. The store's owner took an interest in helping Williams finish high school, where he excelled. Then the owner helped him apprentice in medicine, at which he also excelled.

Finally, in 1880, the owner helped him apply to the Chicago Medical College at Northwestern University, where he was accepted and where he, once more, excelled. Williams was the first African American student in the program.

In 1883, the new Dr. Williams set up a small practice on Michigan Avenue in Chicago. He also taught anatomy at Northwestern University and worked as a doctor for the City Railway Company and, later, the Protestant Orphan Asylum. He was one of just four African American doctors in Chicago at the time and yet his abilities were so obvious that in 1889, just six years into his career, he was appointed to the Illinois Board of Health. Williams wanted more. He wanted to do something more for the city and himself. He was aware that African Americans in Chicago often received poor care from white physicians and nurses. He also watched as African American doctors and nurses struggled to get training and positions, due to racism in hospitals and universities. The challenges facing young African Americans were not waning. At just this moment, a man Williams knew and respected, the Reverend Louis H. Reynolds, came to Williams asking for his help. Emma Reynolds, the reverend's sister, had recently applied to various Chicago hospitals to train as a nurse (she was the first African American to attempt to do so), but she was refused by every hospital because of her race. Her story moved Williams. After discussions with the Reverend Reynolds and other community members, Williams decided there was only one thing he could do: he would open a hospital.[2] At that hospital, he would train African American nurses.

The hospital would come to be called the Provident Hospital and Training Association. It was a bold dream, one in which Williams persuaded other doctors, white and black, and even donors to believe in. Donations came from many sources, including both Frederick Douglass and the Armour meatpacking company (which would also supply the hospital with many patients due to injuries workers incurred on the job). In 1891, Williams signed the lease on a three-story, twelve-room redbrick house at the corner of Twenty-Ninth and Dearborn. Its living room was turned into a waiting room, and a small bedroom at the end of a hall would serve as a surgery ward. In its first year, this makeshift hospital trained seven nurses, one of whom was Emma Reynolds.[3] It also treated hundreds of patients.

Nothing was ever easy at Provident Hospital, but the doctors and nurses made do with what they had. They had to improvise, because of a lack of supplies and the fact that, more than other Chicago hospitals, they dealt with a large number of trauma patients. Everything was difficult, but Williams and his team persevered. His was a story of a hardworking man who overcame and the hardworking nurses who helped him.

But elsewhere in the city, events were conspiring to change Williams's story. James Cornish worked as an expressman, a person charged with the care of packages on trains. The job was a good one, but July 9, 1893, was a bad day. The heat left him soaked with sweat, from morning until six. Worse, the heat did not fade, not even when the sun set. It was the kind of heat that called for a whiskey, which is just what Cornish proceeded to order that night at his favorite saloon. While others in Chicago sampled the best of the world at the White City, as the World's Fair had come to be called, Cornish settled in across town from the fair, among friends. He got his whiskey, took a sip, cracked a flirty joke to the waitress, and walked over to play poker with two friends who were already seated. He felt lucky. A song called "Daisy Bell" was playing loudly from the jukebox. He bounced a little as he walked, eager to laugh, wager, needle his friends, and laugh some more. Then things changed irrevocably.[4]

The sounds around Cornish grew louder. Noise rose like dust. A fight had started. A chair was smashed over the bar. Punches began to land against sweat-damp bodies. Cornish stood on his toes to watch, and then suddenly he was in the scrum. A knife appeared. The man with the knife lunged toward Cornish and stabbed him in the chest. The man pulled the knife back out, someone screamed, the crowd dispersed, then sirens started and several women bent toward Cornish's body, which now lay on the ground.

An hour or so later, at Provident Hospital, Cornish was laid out on a stretcher. His clothes were soaked with blood. He was wheeled into an operating room, where the nurses and Daniel Hale Williams gathered around him. To Williams, Cornish's wound, about an inch in diameter, looked as though it might be superficial. But its location, just to the left of the breastbone, was worrisome. Without x-rays (they were to be discovered two years later, in 1895),[5] there was no way of knowing how deep the wound might be or whether it had reached the heart. The only diagnostics available to Williams were ancient ones. He could feel Cornish's pulse. He could listen to his breathing. He could also put his head or, if he could afford one, a wooden stethoscope to Cornish's naked chest and listen for its wild sounds.[6]

Initially, apart from the hole in his chest, Cornish seemed okay. His pulse was normal. His heart beat. He was cleaned up, sewn shut, and left to rest overnight. Cornish slept in a bedroom with a window that looked out across the city. He had not yet had a chance to inspect his surroundings. He was too weak and then too tired. Warm air blew through the curtains over him. Within hours, his condition, which had seemed stable, began to deteriorate. Dr. Williams was called back in. He ran to the room and up to Cornish's side, where he put his ear to his chest. Cornish's heartbeat was weak, and then, as Williams listened, it seemed to disappear entirely. The heart was still beating, but faintly. On July 10, Williams concluded that the knife must have penetrated more deeply than he had initially thought—all the way into the heart.

A knife to the heart can wreak havoc, though the precise sort of

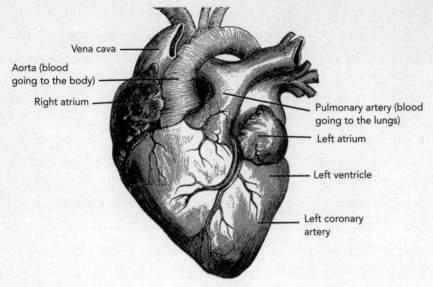

Vena cava

Aorta (blood
going to the body)

Right atrium

Pulmonary artery (blood
going to the lungs)

Left atrium

Left ventricle

Left coronary
artery

The heart, with key parts of its geography indicated. *(Courtesy of ilbusca/
Getty Images)*

havoc depends on the details of where and how the knife enters.
The heart has two sets of pumps. Together, the left atrium (LA in
the figure) and left ventricle (LV) make up one; the right atrium
(RA) and right ventricle (RV) the other. Each atrium (from Latin
for "hall or court, a gathering place") sits atop its corresponding
ventricle. When the left atrium contracts, it gently squeezes blood
into the left ventricle. The blood does not need much of a push, as
it is moving from an area of high pressure into one of low. All it needs
is a little nudge. The left ventricle then contracts much more force-
fully, sending blood throughout the entire body, down the arteries,
to the arterioles, and then through the six hundred million capillaries, each tube of which is just a single cell wide. The force of the
left ventricle's contraction would be sufficient to push water five
feet up into the air or, as is the need in the body, to push blood
through the more than sixty thousand miles of blood vessels in the
human body.

At the same time that the left atrium and then left ventricle contract, something similar happens in the right atrium and then right ventricle, except with less force because the blood leaving the right ventricle does not need to go through the whole body. It needs only to find its way to the lungs, where capillaries rest on three hundred million air sacs,[7] and hemoglobin, in red blood cells in the blood, releases carbon dioxide and gathers oxygen.

The sounds of the heart, at least the most conspicuous sounds, are those of the valves between the atria and ventricles (the mitral on the left; the tricuspid on the right) closing when the ventricles contract (and, in doing so, preventing blood from flowing back into the atria) and then, more loudly, the valves between the ventricles and the arteries (the aortic on the left, the pulmonary on the right) closing once the ventricles have finished contracting (which prevents blood from flowing back into the ventricles): *lub*-dup, *lub*-dup. The sound of the heart is the closing of these valves, day in, day out, billions of times in a fortunate human life.

So much depends upon the heart's pumps. The blood that is pumped out of the left ventricle travels into the aorta, which serves as a superhighway from which blood is shunted off into branches to the arms and brain, to the internal organs (intestines, liver, kidneys), and to the legs and genitals. Meanwhile, the right atrium and ventricle receive the blood that has come back in a different form than it went out—now the blood is depleted of oxygen and full of carbon dioxide. This "used" blood is pumped to the lungs (via the pulmonary circulation; *pulmo*- comes from the Latin for "lung"), where blood cells, in effect, exhale carbon dioxide and inhale oxygen. The oxygenated blood then flows to the left atrium, where the process begins again.

All of this is happening in you right now. It happens in waves: contraction, relaxation. The contraction is referred to as *systole* (from the Greek for "to pull together"), the relaxation, *diastole* (from the Greek for "to separate"). Hold your hand to your neck, and you can feel, in the expansion and relaxation of your carotid arteries (which supply your brain with oxygenated blood), the consequence of your heart's pumping.

That is what you hope for, anyway, but when Williams felt his patient's neck, that is not what he found. The assault on Cornish's internal machine had made the heart both weak and slow, and the pulse could barely be felt. A knife wound can provide a new hole through which blood pours into the body cavity instead of into arteries. It can also — and this is far worse — interrupt the ability of the heart to contract.

Just what was happening in Cornish's body was hard to say. Today we would have many more clues than Williams had. We could look at an x-ray, a sonogram, a CT scan, or an MRI. A catheter might be threaded into a patient's heart to release dye that would reveal, in the x-ray, the location of the damage. A machine would record the rhythm of the heart. What we would know today would not be perfect, but it would be useful. Williams had virtually nothing except the weakening of Cornish's heartbeat and his obviously deteriorating condition.

The weakening of a patient's heartbeat might be due to a problem in the heart itself, but it might also be due to loss of blood, to which, we now know, the body can partially respond. The arteries in our bodies are muscular. They contain a layer of smooth muscle. Smooth muscle is not under our conscious control, but it is under our bodies' unconscious, autonomic control. The muscles in our arteries do not push blood along — that is the heart's unique role — but they can widen or narrow the vessels to slow or speed up its passage. And one sort of artery, the arteriole, can actually stop the flow of blood. Arterioles are the narrowest arteries — they meet up with the capillaries, which then connect to venules, which in turn connect to the veins that carry the oxygen-depleted blood back to the heart — and arterioles are narrow enough that when they contract, they close. They do so to influence the flow of blood in the body. When your fingers are cold, blame the arterioles, but also thank them because they are, based on the condition of your body, helping to move blood where it is most needed.

If Cornish was losing blood, the arterioles would have begun to shut off the flow through nearly all of the capillaries in the body (except those in the three organs that never, except in the very worst circumstances, lose their blood flow: the brain, the heart, and the lungs). When this happens, the pulse weakens, the extremities grow cold, and the body struggles to preserve that which it cannot do without.

With his patient deteriorating, Williams had to make a decision. He knew Cornish's heart was broken, but he was at a loss to say precisely how or why. No matter the cause, the most likely scenario seemed to be that Cornish, friend to many, son to one good mother, was about to die.

Knife wounds to the heart were remarkably common in 1893. They remain common today, though they are now rarely fatal. If you are stabbed in the heart, raced to the hospital, and operated on, you stand about an 80 percent chance of survival. A trauma to the heart can be operated on in any of a variety of ways, or not operated on at all, depending on the condition of the heart. The odds are now good for victims of stabbings, thanks to both technology and the learned skills of surgeons. But in 1893, the most likely consequence of a stab wound to the heart was death. Once the heart started to bleed, whether from a stab wound or some other assault, a patient depended purely on fate to survive, a kind of cardiac destiny. Sometimes the body was able to restrict blood to the core and heal the wound before too much blood was lost. More often, it couldn't. Infections took over, or the heart lost its rhythm. Doctors sought medicines that might cure such wounds, but they sought in vain. And no doctor in the world was known to have successfully operated on a heart, wounded or otherwise. No one, as far as Williams knew, had even tried. It was the Mount Everest of the body, the great mountain not yet climbed. Yet, if Williams was anything, he was the kind of man who tried, the kind who might scale a mountain to save someone. He had tried working on shoes as a young man. He had tried working in a barbershop. He had even tried music and law. He had tried surgery and running a hospital. Now, on July 10, one day after Cornish was stabbed, he would try something even more novel.

Williams and the nurses looked down on Cornish. They all bent over him to closely inspect the damage. It seemed likely that his heart—that bloody engine—was torn, though even that was not entirely certain. If it was torn, Cornish would die from internal bleeding or, depending on the severity of his wound, heart failure. Williams could do what every other doctor in the same situation had done for the past ten thousand years, which was walk away. Or he could operate. Whatever he did, the heart was there, just inches from his face as he bent over his patient, just under the surface and yet for all of time so very far away.

One can imagine the sort of person it takes to perform the first surgery ever on a heart. He or she would need to be self-confident but also eager to go beyond what had been done, both to save a patient and to advance humanity. Williams was such an individual. On July 10, 1893, the operation began. Williams was handed a scalpel and the other tools necessary to cut into Cornish. He was about to attempt a feat surgeons all over the world had advised was too dangerous and immoral. Success or failure, Williams was about to make history.

The human heart beats, on average, about a hundred thousand times a day, pumping 7,500 liters of blood through arteries and veins. But this was no average day. On this day, Williams's heart would have rabbited along, pushing extra oxygen to his eager brain. Six other doctors had also gathered in the room. Williams swore he could hear their hearts too. This is the great irony of surgery and, more generally, medicine: that a doctor in one body bends to mend a patient in another body, the doctor relying on the same parts (her heart, her brain, her skin and flesh) she aims to fix in her patient. The room was more than a hundred degrees, and even before Williams began, everyone was sweating. Now, with anxiety and adrenaline, they were dripping so much that the floor was wet. Williams wiped his head and then, with the nurses at his side, inserted the blade into Cornish's wound and cut a six-inch incision. He inserted his right hand through the incision and pulled one of the ribs away from the sternum to make a hole, a kind of window through which he could look at Cornish's heart. He siphoned away the excess blood and, for the first time, had a clear view of the heart. In general terms, it was an ordinary heart, somewhat larger than a clenched fist, about five inches long, three and a half inches wide, and two inches thick. What was not ordinary was that it lay bare, as naked as a heart can be, suddenly at the mercy of insight, skill, and luck.

The atria and ventricles of the heart are surrounded by the pericardium (the word comes from *peri-*, Greek for "around," and *cardia*, Greek for "heart"), a smooth, oily sac. As Williams looked at Cornish's pericardium, he could see where the knife had gone in, through the pericardium and into the heart muscle. Williams had very little time to decide what to do. It was too late to turn back. As he looked at the heart muscle, it seemed as though the wound had sealed itself over, closed with the pressure of the contractions of the heart. Right or wrong, this observation, along with perhaps a hiccup of trepidation, led him to focus on the pericardium. He would not be the first to operate on the heart muscle, but he would be the first to sew the pericardium. He cleaned the wound as best as he could (antiseptics were new and one of the reasons Williams had a chance, albeit relatively slight, of preventing infection) and then began to sew with catgut thread. The needle sank through the pericardium and then, with a tug, came back out the other side. It sank again. As it did, the heart beat, though only weakly. Williams tried to time his efforts with the heartbeat. The hope was that the stitched-together pericardium, however flimsy the sewing, would stabilize the heart. When he was finished, Williams took a deep breath and stepped back to inspect his work. Without meaning to, he beamed a little. Time would tell if Cornish would live to beam back, but whether the patient lived or died, Williams had just changed the trajectory of medicine. He had taken the plunge into the heart. Others would follow. They would not be able to resist the temptation to raise their scalpels and, one heart after another, cut.

Williams worked a little over a hundred years ago, just yesterday in the context of the human story. The history of surgery is ancient. Stone Age needles were once used to suture cuts in preagricultural Africa. Army ants were used to close wounds in India and the Americas (the ant bit down on the wound and its jaws locked tight; one ant for small lacerations, two for big ones). As societies became larger and more sophisticated, the surgical repertoire expanded. With the birth of agriculture came civilization, writing, and systematic attempts to create new forms of medicine. In ancient Mesopotamia, China, and elsewhere surgeries were attempted on many different parts of the body, even the brain. As early as eight thousand years ago, medicine men chanted, burned herbs, and then drilled holes into people's skulls to "relieve pressure" (at one site in France, dated to 6500 BC, one-third of skulls showed evidence of drilling). Many of these surgeries were successful, or at the very least not fatal. Amputations were also done, as were removals of stones from bladders. With time and a kind of mortal inevitability, more and more parts of the body came to be operated on, until, at the time of Cornish's incident—roughly eight thousand years into the history of surgery—someone somewhere had either effectively or experimentally (or both) operated on nearly every single part of the body. The brain, eyes, arms, legs, and stomach had all been cut and sewn, but not the heart.

The heart was special. Before 1893, for the thousands of years during which humans practiced medicine, the heart was viewed as either functionally or philosophically untouchable. The standard medical text in Williams's office (a converted bedroom closet) offered this: "Surgery of the heart has probably reached the limits set by Nature to all surgery; no new method, and no new discovery, can overcome the natural difficulties that attend a wound of the heart." Any doctor who dared operate on the heart would be shunned and, many thought, should be. Theodor Billroth, a dominant force in European surgery at the time, argued that a surgeon who tried to suture a heart wound deserved to lose the esteem of his colleagues. Williams had crossed the last anatomical frontier.

Several factors contributed to the perception of the heart as inviolable. Many cultures had long viewed the heart as the source of emotion, the mind, and the soul. Such sentiments persisted in the late 1800s. The French surgeon Ambroise Paré gave them voice when he wrote, "The heart is the chief mansion of the soul, the organ of vital faculty, the beginning of life, and the fountain of the vital spirits...the first to live and the last to die." The modern Valentine's Day link between the heart and love relate to these ancient concepts echoed by poets across centuries in both their words and their deeds. Take the death of Percy Bysshe Shelley. Shelley was cremated, but, according to his friends, his heart did not burn, so powerful was its poetry. While doctors practicing in the late 1800s had a less mystical interpretation of the heart's function than Shelley's friends, they still imbued the heart with a kind of unknowable magic, the sort we now seem to reserve for the brain. Who could really say what lurked in its dark caves? If not Sirens and Fates, it held at the very least the essence of life.

The taboo associated with operating on the heart deterred many doctors. But if that were the only problem, some bold surgeon would have violated it long before Cornish ended up on the table. The field of surgery has long attracted and trained (albeit not exclusively) aggressive, overconfident individuals who do what seems impossible rather than what is allowed. The real challenges were technical problems in the art and science of surgery. The heart beats. It is the most lively part of the body, wild and hopping, so any operation would have to be done in time with the beating, as though in a sort of dance, a surgical waltz. Antibiotics had not yet been discovered, so the odds of infection were high. Nor did x-rays exist (much less angiograms and CT scans), so no one could see what was wrong with the heart until the chest was opened. Then there was the issue of breathing. No machines existed for keeping airways open during surgery. For all of these reasons and more, every time someone with a bullet or knife wound to the heart came into a hospital anywhere in the world, the only option was to keep an eye on the patient and watch as the body healed itself or, as was often the case, did not.

Thirteen days after his surgery, Cornish, who was still in the hospital, had his fate announced to the world. He had survived. In the newspaper articles that followed, Cornish was described as a fortunate soul, Williams as a hero. Williams was heralded as the first surgeon to have operated on the human heart, and successfully at that. Williams was not modest about the procedure. He would go on to do others and even brag when he did; as he would say of himself in a newspaper article, "Successes crowned [his] attempts in nearly every case." Meanwhile, Cornish was still in the hospital, where, suddenly, on August 2, he got worse.

Williams rushed to Cornish's bedside. Cornish's blood pressure had dropped dramatically, but Williams was not sure what was going on. With his recent success at his back, he decided to open Cornish up again and conduct the second heart surgery in the world. He made a new incision, undid the original stitches, drained the space between the pericardium and the heart muscle, sewed the pericardium shut again, and then stepped back. It was, he thought, almost easy to work on the heart. Cornish left the hospital on August 30, 1893, alive.[8]

Cornish went home to his family and lived a long and largely happy life (one exception to that general happiness being a return to Provident Hospital with a head wound from another bar fight). He died in 1931, thirty-eight years after being stabbed in the heart. But the broader consequences lasted far longer than Cornish. Williams broke the barrier to the heart, the myocardial ceiling. Once surgery had been done on the heart, others began to operate too, and it would be Williams's model, as well as broader changes in medicine, that built the first step that led to modern cardiology. We think of the medicine of hearts as well established, but the truth is that every medical treatment of the heart, as well as most of what we know about the organ, has come since 1893. In that year, one heart was operated on, twice. In 2010, more than half a million hearts were operated on in the United States alone.

The character in this story who really started everything, that man who stabbed Cornish in the bar, is forgotten. He could not have anticipated the series of events that would transpire thanks to his knife. To paraphrase Dr. Harry M. Sherman, speaking at the annual meeting of the American Medical Association in 1902, the road to the heart is only two or three centimeters in a direct line, but it had taken surgery nearly ten thousand years and a bar fight to travel it. Meanwhile, time and perspective have modified our understanding of the events surrounding the surgery Williams performed. That such a major advance was made in a poor hospital by an African American doctor and African American nurses just thirty-one years after the Emancipation Proclamation is astonishing. We tend to regard technology as the source of many innovations, and yet Williams's advance was something different, progress through some combination of hubris, intellect, and will. He and the doctors and nurses he gathered around him had the necessary mix of wherewithal and confidence to try, and the skills to carry through.

Time has also added more context to the question of whether Williams was really the very first to perform a heart surgery. He thought he was, but he had actually been preceded by two years. In Alabama in September of 1891, another doctor, Henry C. Dalton, had performed a remarkably similar surgery (again on the victim of a stabbing), though the news would not be published until two years after Williams operated on Cornish.[9] Williams's prominent surgery was the one that made doctors aware of what was possible with a knife, a sewing needle, and some catgut.

We might hope that what motivated surgeons such as Williams to do new procedures was their goodwill toward humanity. There was some of that, but there was also the same motivation that drew Mallory up Everest: Mallory climbed Everest "because it [was] there." Like Everest, the heart was there. The next step up the mountain was to actually cut into the muscle of the heart.

On September 9, 1896, a gardener arrived in the Frankfurt am Main hospital with his clothes soaked in blood, but once cleaned up, he seemed to be stable—until suddenly he was not. With the patient's health quickly worsening, the surgeon Ludwig Rehn was brought in. It seemed as though the gardener would die. There was nothing more to do, a situation that emboldened Rehn. Rehn decided to cut into the gardener. He opened the man's ribs and saw the heart. It was beating beneath a sea of blood, pumping and spitting. Rehn pushed his finger into the heart and found a hole. The feeling was marvelous. The heart slipped beneath his finger as it beat. He was amazed to find that it was strong rather than weak, as he had assumed it would be. He held his finger in the hole as best he could and then, seizing the moment (and a needle and thread), began to sew, one stitch for each beat. Rehn, like Williams, was successful[10] and hopeful. As he wrote of the day, "This proves the feasibility of cardiac suture repair without a doubt! I hope this will lead to more investigation regarding surgery of the heart. This may save many lives."

It did. In 1907, Rehn reported that 120 surgeries on the heart had been performed around the world, about 40 percent of which had been successful. The results weren't perfect—they still are not, even though current mortality from the same surgery is just 19 percent—but they sure beat the near certain death that had previously been the outcome of stab wounds to the heart.

Before 1893, the heart was simply not touched. Beginning in 1893, it was touched and, surgery by surgery, more effectively sewn back together when damaged. There was a sense of progress even when, in retrospect, progress seems to have been slow. In 1923, Dr. Walter Lilienthal of Cornell Medical School noted in *Time* magazine that there had been major successes in heart surgery, and he went on to list inventions that today seem modest—a phonograph to record the sounds coming from stethoscopes, a camera set up to take pictures of the moving heart, the realization that adrenaline could speed up the heart (it had recently been injected into the heart of a seemingly dead baby boy and had saved him).[11] Yet at the time, this all seemed like immense advancement, advancement that would only accelerate. There are many stories to tell about the years that separate Williams and Rehn from modern medicine, stories of ambitious individuals who believed they could conquer our most tempestuous organ in new ways, and of patients, be they presidents or paupers, who lived or did not as a consequence. Technical progress was made, sometimes to the benefit of patients, sometimes at their expense. Hearts were stopped and started again. Hearts were even moved, beating, from one human to another until such surgeries were perfected and came to seem, if not quite ordinary, at least mechanical. But before we consider the story of the past hundred years, let us begin with the story of the prior thousand.

2

The Prince of the Heart

It was an unbelievable moment. As spectators looked on, a man gutted a Barbary macaque in the middle of a city plaza. There, before the crowd, he challenged everyone to put it back together. The moment was madness, but the man was not mad. He was, almost inarguably, the most important medical scientist in history. His name was Galen, and this moment before the crowd was his coming-out as a performer and a scientist.

Galen, or Galenus of Pergamum, was born in AD 129 in Pergamum, near the Aegean coast of what is now Turkey. He was, at least by his own reports, a kind son and a dutiful student. Upon finishing school, and at the suggestion of his father, he traveled to Alexandria, the great city of learning. Galen's father had seen in a dream that his son would be a great doctor. Galen would later speak often of this dream,[1] emboldened by his father's premonition to strive even more eagerly. Following his training and his father's death, Galen looked to establish himself. He needed a livelihood. After the sort of wayward traveling that even today makes parents nervous, at the age of twenty-eight, he decided that he would like to begin his career as a medic to gladiators, once again in his hometown of Pergamum. There was a problem, though. He needed to be chosen from among many candidates. This was too important an opportunity to leave to chance.

The potential gladiatorial doctors were asked to meet in a

GALENVS. 19.

Explicat Hippocratem, nec latum præterit vnguem;
Eloquio sectas, Thessalicosque mouet.
Nulla est cognitio, non experientia rerum,
Quam non attigerit, perpoliítque simul.

E

Galen, the father of modern medicine, whose influence has yet to fully wane.
(Courtesy of the National Library of Medicine)

public area. Galen is said to have brought with him a very hairy and
ill-fated ape (actually a Barbary macaque). While the other doctors
looked on, Galen eviscerated the macaque. This was his moment. It
was madness; it was horrible. But Galen had his reasons. Standing
above the animal, he challenged the men around him to put it back

27

together again. No one but Galen could. He got the job (or, in some tellings, at least secured the job that was already his). Galen might have been trying to say that in order to tend to gladiators, you needed to be able to put their guts back together. But the message the other doctors probably heard was something more like *I'm so crazy, I'll cut open a macaque; you don't want to try to take this job from me.* Either way, it worked. Galen had already acquired some of the brash skill that would bring him fame.

Galen took up his position with the gladiators. He traveled with them during the winter, spring, and fall training months; he worked alongside them. He was a fight doctor, and his was a world of sweat and blood and well-trained men, men whose bodies and, particularly, hearts worked slightly differently than the norm. We know now that in an endurance athlete, the heart's left and right ventricles expand in order to pump larger volumes of blood to the body. The relaxation of the heart between beats also becomes more extreme. By contrast, in a strength athlete, such as a bodybuilder, the heart's ventricles do not necessarily get bigger, but they get stronger, and the heart relaxes less, rather than more. In the movies, gladiators look like bodybuilders. In reality, they probably looked more like chubby, small-town strongmen. They ate a special vegetarian diet of barley and fava beans to become somewhat fat, the idea being that the fat might help protect them from wounds. Yet, despite the fat, gladiators did exercise, and it seems likely that their bodies were the result of a mix of strength and endurance. One expects that their hearts were a mix, too—strong and big and never terribly relaxed.

In the summer months, fights were staged, and Galen would wait for the injured. A coliseum rose above him, and in its stands Galen could hear twenty-five thousand fans cheering and booing. They loved the spectacle of the gladiators battling in the dry dirt. The fans felt as though they were fighting too. Galen heard them cursing. He heard them moving. He heard their bodies above him, around him, a great mass of hands and legs, flesh, and, buried in the bodies—hidden but so easily revealed—hearts, livers, kidneys, veins, and arteries, all of which he knew were there and yet could not explain. As he stood in the heat of a crowd, he dreamed of greatness, but not that of the gladiators. He dreamed of his own.

The gladiatorial battles alongside which Galen stood were the precursors of all sporting events that would follow—every one you have ever been to. It is not hard to see the fans of gladiator events in the faces and actions of soccer or football hooligans. And the gladiatorial arena would be the precursor to the stadium, but also to the surgical theater. To Galen, the gladiators were titillating as they ran at each other with their weapons raised. But more titillating was the subsequent struggle, his struggle, to save the men who had been wounded. Anyone could kill a man. Only Galen could so consistently bring a man fated for death back to life, at least according to him. He wanted and felt he deserved a crowd.

Galen's predecessor had lost many gladiators to injury. The wounds were too deep, infection too overwhelming. Gladiators died one after the other. But not on Galen's watch. In the entirety of his time tending to the gladiators, just five men died.[2] Perhaps being able to stitch up a macaque actually was useful in stitching up gladiators, or perhaps Galen was ambitious enough to be both self-aggrandizing and great.

As Galen sewed the bodies of gladiators back together, he made scientific discoveries. The bodies of the gladiators were specimens in which the muscles, nerves, and veins were, if not exactly obvious, then at least more obvious than in the average man. The gladiators could be learned from. Seventeen hundred years before Leonardo da Vinci would carefully sketch out the body's external details, Galen was gaining daily experience with its internal ones. The gladiators' wounds were, as Galen would write, "windows into the body." It was thrilling for him to look into wounds. Was it joy, a kind of love? Galen would later do experiments in which he would track the heartbeats of lovers when they were reunited (or pulled apart). Love made the heart pound, and Galen loved discovery; it made his own heart race beyond his control. Today we know that love, rage, and other strong emotions affect the amygdala, a group of neurons in the most ancient part of the brain. The amygdala triggers the release of hormones that affect many organs, including the heart; the hormones can cause the heart to speed up, which sends more oxygen to the brain. All Galen knew was that he could feel this effect. He could feel his heart running as he saw the living, working parts of bodies, parts very few other people had ever seen.

In Galen's Roman Empire, most doctors would not have seen the heart even in a dead body. The Romans waffled on what was necessary in the afterlife and so prohibited all human dissections; better to be safe than sorry. The sliced-open bodies of the gladiators would have to suffice for Galen. Working on these bodies, Galen sometimes paused longer than he should have, to look. He may even have seen a beating heart. (He definitely saw one later in life, when asked to tend to a boy with a chest infection. Galen cut into the boy's chest and saw his beating heart. He may even have cut into the boy's pericardium, two thousand years before the next such cut, by Daniel Hale Williams in Chicago.) Certainly, Galen saw the roads of arteries and veins. He saw enough to begin to sketch, first in his mind, then on a papyrus, a semblance of the intimate geography inside all of us. He was the first real geographer of our untraveled reaches—Captain Cook of the high seas of blood— and though he would err in how some of the regions connected, mistake some peninsulas for islands and so forth, his would be the map that would allow all of those who followed to proceed, to hesitantly check the boundaries he had so carefully limned.

As Galen continued in his role with the gladiators, he learned and profited, but the former more than the latter. He wanted more—more money, fame, and understanding. Eventually, Galen retired from his gladiator job and began to work as a sort of traveling doctor and showman. Doctors already existed, but their treatments had very little to do with actual diagnoses; Galen was, arguably, the first doctor who aimed to understand what caused ailments and then to treat those causes based on the results of testing the treatments on multiple patients. He had learned about science from Alexandria. Now he was learning to treat illnesses (not just mending wounds, which was his focus with the gladiators) by empirical trial and error. His doctoring allowed him to see more kinds of afflictions. His performances, at which he would dissect an animal, treat a patient in public, or otherwise make a spectacle, allowed him to both garner support and, however informally, teach. His research, which often occurred during public dissections and displays of doctoring ability, allowed him, each time he looked, to understand more. It was the performance and the research that motivated him; he wanted to understand and then be able to display that understanding to the masses.

During these years (and, for that matter, in the fourteen hundred years to follow), Galen had no rivals. Across the entirety of the Roman Empire, stretching from modern Scotland to Egypt, Galen's accomplishments were lauded. He became a legend in his own life, so much so that at the margins of the empire, the story of his life became exaggerated, imbuing him with the status of a sort of half-god. His reputation traveled by word of mouth—in response to his performances, for example—but also via his writings. Galen wrote volume after volume about both what he had discovered himself and what was already known but not well consolidated. These volumes were greedily consumed throughout the Roman Empire and beyond. Thanks to this fame and success, Galen eventually became the physician to the emperors. He tended to their royal bodies with tender effectiveness, just as he had once tended those of the gladiators, but the pay was much better. He also continued to write. Or maybe it is better to say he continued to speak. He wrote his books by talking. A dozen very busy scribes recorded his every word, words that eventually came to comment, in great and lasting detail, upon the biology of the heart.

When Galen recorded what he knew about the heart, he built on millennia of observations, some of them formal, others part of the sort of everyday understanding of the organ every hunter must have seen beat when he dissected the freshly killed body of his prey. All around the world, hearts had been seen. After an elephant was killed, its ponderous heart was thrown to the ground, a giant house of muscle to be entered through the blood vessels' wide caves. The tiny hearts of birds were dried and worn like decoration on strings. Native Alaskans stood beside whale hearts and felt small. One ten-thousand-year-old cave painting in Pindal, near Altamira in Spain, shows a bright red heart inside the body of a mammoth. The hearts of animals were varied and yet recognizable as hearts. They beat the way human hearts would ultimately be seen—in war and accidents—to beat. Even before we knew what the heart did, we knew that it was the measure of a life, be it in a bird, a squirrel, or one's kin. It sped up in fear, eagerness, or bravery, and when it stopped, the creature that housed it would die. The heart was and is the most deadly place to be stabbed, the most vulnerable lump of muscle, only marginally protected beneath the shallow cage of the ribs. The modern wounds from knife fights are still very often in the heart. Whatever it was, in whatever body it might be found, the heart was simultaneously weak and powerful.[3]

Just how the powers of the heart were understood varied from place to place. Yet those stories of the heart we know the most about, despite being separated by oceans and time, share many similarities. To the Aztecs, for example, the heart was infused with the sun's borrowed fire, but that fire had to be returned. To return a bit of fire to the sun, the Aztecs cut the beating hearts of sacrifice victims out of their bodies. The priests who did this cutting (and, as if it weren't already brutal enough, tearing) would have seen more living hearts than anyone else up to that point in history. They would have known the weight of a heart and many of its specific details. They gathered enormous clay jars full of human hearts that would, at the end of the season, be poured into water, whether cenote or sea, in gratitude for the sun-blessed crops. In the intervening months, the hearts would be placed where they could be observed and contemplated. The result of that contemplation was not recorded; the Aztecs left no comment about what they thought the heart did or why—only that they, like nearly every other culture, regarded it as important. The Aztecs decided to remove hearts, not livers, kidneys, or stomachs, from those who were sacrificed.

Two continents away, the Egyptians left the hearts alone in the bodies of mummies to be transported into the afterlife, when the heart's fire remained necessary. Later, in Greece, Plato also linked the heart and flames. He wrote, "The swelling of the heart which makes it throb with suspense or anger was due to fire."

In most cases, the specialness of the heart extended beyond mere biology. Across cultures, the heart is the most frequent seat of the soul, the spirit, or the breath of God or the gods. Among Christians, Jesus was said to live inside the chambers of the heart. In ancient Egypt, the heart was the home of both the soul and, by some reckonings, consciousness. There are exceptions, of course; one Australian tribe believed the soul to live in the fat around the kidneys,[4] and the Mesopotamians found their souls in their livers, but these cases stand out for their very unusualness.

Explanation of just what the human heart was awaited the origin of scientific societies. Long before Galen, the first detailed account of the heart's biology is found in the Ebers Papyrus, an Egyptian medical encyclopedia recorded on a sixty-foot-long scroll written by the scholar Imhotep in roughly 2600 BC. The oldest remaining copy is a relatively recent one from 1700 BC. If one takes the time to carefully unroll and read it, several ancient stories of the heart are revealed. One can be found in a section entitled "The Beginning of the Physician's

Secret: Knowledge of the Heart's Movement and Knowledge of the Heart." There, Imhotep wrote that there are vessels "to [the heart] from every limb....When any physician...applies the hands or fingers to the head, to the back of the head, to the hands, to the place of the stomach, to the arms or to the feet, then he examines the heart, because all his limbs possess its vessels, that is: it speaks out of the vessels of every limb." In the Ebers Papyrus, the heart is considered to include the vessels leading to and away from it; the heart is the whole cardiovascular system, the muscle itself but also the rivers coursing from head to toe and, though it was not yet known, back to the heart.

The heart spoke to the Egyptians in terms of both its physical form and its metaphysical possibilities.[5] Yet the Egyptians, for all their knowledge, seem to have understood very little about what the heart had to say. Its movement conveyed meaning that could be felt anywhere in the body. It beat out a story, but what was the heart carrying on about? The Egyptians could not yet offer a compelling translation of the heart's rivers and backwaters, their murmuring synonymous with being alive.

Finally, in Alexandria, Egypt, twenty-three hundred years after Imhotep, the science of the body began to advance anew, as did nearly all other intellectual endeavors. In 330 BC, Alexander the Great had founded Alexandria. It was to be an ideal city, ruled by Ptolemy I. In Alexandria, life on earth mattered and affluence abounded, so philosopher-scientists had both the mandate and the funds to begin to explore the material world—and that included the body. The science of Alexandria began in a new and grand museum called, simply, the Museum, or the Alexandrian Museum, a sort of university dedicated to the intellectual muses. Down the street, the Library at Alexandria stored as complete a record of the history of the world as had ever existed.

Walking through Alexandria, one could bump into Euclid absent-mindedly pondering his new math or Eratosthenes trying to measure the diameter of Earth (and coming within fifty miles of getting it exactly right). Then there was Hipparchus cataloging the stars. Hero was at work designing a steam engine. Archimedes came to visit and learn.

The anatomists of Alexandria worked in the museum and read in the library, but their place of real discovery was the medical school. In that school, the earliest of its kind, dissections and vivisections were permitted for the first time in thousands of years or maybe, at least for scientific ends, ever. Criminals were examined while still alive. Some of what we know about our own bodies today is thanks to their horrible fates. Their vivisection allowed philosophers clear views of living human bodies, clearer than would be seen again for two thousand years. Philosophers could hypothesize as to what the body did and then test those ideas, one life at a time.

As they looked into bodies, the philosopher-anatomists of Alexandria built on the ancient knowledge of Imhotep. They had also benefited from recent discoveries. In roughly 500 BC, Alcmaeon of Croton noticed that when he was looking at dissections of animals, the arteries and veins (though he didn't know what they did at the time) seemed "different" from each other, though precisely how or why was beyond what he was prepared to say. Presumably others had noticed these differences too, but Alcmaeon recorded what he saw and got the credit. Aristotle (384–322 BC), adviser to Alexander the Great, built upon Alcmaeon's observations and made a few new ones. Aristotle looked carefully enough at the heart to name its parts. He thought he saw in the heart three chambers, the left ventricle and atrium and the right "chamber," which we now regard as being composed of two parts, the right ventricle and atrium. Aristotle reaffirmed too the importance of the heart, bestowing it with the soul (as others had), but also with thought itself. To Aristotle, the brain was filled with nothing more than mucus;[6] the heart, however, was a thinking man's organ. Today we describe ourselves as thinking with our brains, and it is hard to imagine any other repository of our thought, and yet for much of history, the location of the human mind drifted in the body, subject to new theories.

In Alexandria, Herophilus (335–280 BC) built upon this knowledge and was the first to notice that one type of vessel (what we know today as arteries) was thicker than the other (veins), and muscular. For that he was lauded.[7] Even in Alexandria, progress was painfully slow when measured against the extent of the ignorance of the time; the ancient boats of discovery were still bumping clumsily along the body's uncharted shores. Herophilus (who is sometimes called Chalcedon in reference to his place of birth) also made another discovery. He thought that he had discovered that both arteries and veins contained blood. Until then, the arteries and veins, as well as the heart itself, were thought to be filled with air (the word *artery* is even from the Latin word meaning "air ducts"). This misconception arose because in death, without the pressure of a beating heart, blood quickly drains out of the body's arteries (and, to a lesser extent, veins). In the dead, the arteries are air ducts, the heart itself a vessel. So persistent was the idea that the heart was filled with air that Herophilus's friends and his clever colleagues thought he was wrong. Erasistratus (304–250 BC), Herophilus's slightly more youthful contemporary, was among those who insisted that air alone inhabited the heart's muscle, as well as the arteries and veins. (Erasistratus was no dummy; he was the first of the body's explorers to correctly suggest that the heart was a sort of pump.) Herophilus and Erasistratus agreed on one thing: whatever the substance traveling from the heart out through the blood vessels, it invigorated the body with life. It was on the basis of the knowledge of these men that Galen began to build his empire of observations, a kingdom of facts that would stand the test of centuries.

Galen probably knew the history of his intellectual forebears better than we do today, since the great Library of Alexandria burned down in the years after Galen, and we are left with just footnotes of the knowledge that had accumulated by Galen's time. Galen himself went to Alexandria for some of his training. Then and afterward, he was especially interested in the heart and what we now call the cardiovascular system. Galen thought of this system as being mechanical (he seemed to have already banished both the gods and the soul from the body, to his own satisfaction), and in order to test the makings of its mechanism, he needed and wanted to do experiments. But he had a problem: Now that he was no longer mending gladiators, he was rarely able to see inside human bodies. Vivisections had become a thing of the past. They were off-limits to him, as were dissections of humans in general. He could not even do simple experiments on humans, experiments of the sort now permitted—albeit only after approvals are obtained—in Western medicine. He wondered what would happen if you clipped a vein shut. Would the blood pool above it or below it? No one had ever thought to ask the question before. In the absence of human subjects, how could he find out?

Galen's progress came to rely on the law of similarity, namely, that the bodies of different animals are sufficiently similar to one another that if you study one, it will tell you (imperfectly and yet still usefully) about others. Two thousand years before Darwin, Galen recognized and relied upon the kinship of humans to other animals, and we still rely upon it. When new products or treatments need to be tested, they are first tried out on guinea pigs, rats, dogs, cats, or monkeys. They are tested on those animals because the animals are similar enough to us to be useful measures of how our own bodies operate. Then, if everything works in the nonhuman animals, the same products and treatments are tested on humans (college students, fairly often). As Galen put it, "The bodies of different animals are the same so you can study one animal and learn about the other; you can study dogs and learn about humans." Researchers do animal testing because Galen popularized the approach; his legacy lives on in the millions of rats, mice, and guinea pigs used in labs as proxies of human bodies.

Galen did not believe absolutely in the law of similarity. He knew that dogs and macaques were not humans. He knew that similar bodies did not mean identical bodies (this would later be forgotten by his disciples), and yet he thought that if he dissected and experimented on dogs, he might begin to understand the human body. While human dissections were prohibited in ancient Rome, those of other animals were not. Galen could dissect as many dogs as he wanted, so he did. He also dissected pigs, goats, sheep, horses, asses, mules, cows, lynxes, stags, bears, weasels, mice, snakes, fish, birds, and an elephant—along with whatever else he could catch or import.[8]

Galen confirmed that the heart was filled with blood and that the veins and arteries were different from each other, and he observed for the first time that the blood in the veins and arteries was also different. The blood in the arteries was red; that in the veins was purple. One dissection at a time, he was beginning to make the observations on which a modern understanding of the heart and cardiovascular system could be built.

Galen, having mapped the heart's general features, decided the next task was to understand the heart's function. From Galen's classical perspective, each organ had a function and a kind of internal autonomy. He believed that many organs produced vital substances; this was his modest retooling of a more ancient cosmology of the body handed down to him from the early philosopher Hippocrates (born around 460 BC) and his disciples in the form of a series of books called the Hippocratic Corpus. The lungs produced phlegm, the gallbladder bile, the spleen melancholic black bile, and the liver blood. For a human to be healthy, these substances needed to be balanced. In the writings of Hippocrates, the heart, though, was different; it pulled substances to it, and it was possessed of a gravity of the kind that would much later be attributed to the sun. The heart, in essence, demanded from the other organs, and they obliged. In light of these perceived functions, Galen began to craft a more detailed vision of what he thought the cardiovascular system did.

Galen was burdened by the old ideas about organs as he looked at bodies; they colored what he saw. Through the cultural lens of ancient science he came to believe veins originated in the liver. There, he theorized, digested food from the stomach mixed into thick purple blood and traveled through the veins to the rest of the body, where the blood was depleted by the body's demands before traveling to the heart. Once in the heart, the blood traveled two places. Some of it (the vital blood) traveled to the lungs, he (rightly) imagined. But much of the blood traveled directly from the right to the left ventricle, through tiny pores, he thought (incorrectly). The blood from the lungs, he believed (again rightly), traveled back to the left side of the heart, from whence it would distribute the "spirit" it had gathered in the lungs to the body via the arteries. Galen somehow also understood that in some parts of the body, the arteries and veins met.[9]

In the most charitable reading of Galen's understanding of the heart, one could say he essentially discovered the circulation of the blood. He was wrong about the role of the liver (although only partially; the sugar in the blood is released from the liver, so part of what's in the blood does come from the liver). He was wrong, too, about the holes in the heart through which blood flows right to left. But even this error is far from absolute. If we conjecture that some of the bodies that Galen was studying were fetuses, his ideas are even more right than they seem to be. In a mammalian fetus, blood actually passes directly through a hole, called the foramen ovale, that's between the right and left atria (not the ventricles). This hole seals up during development, but it is open for a while. The main problems in Galen's model were his insistence on the ancient belief that blood was produced in the liver and consumed by organs and, of course, his theory of what blood was and did. But he was getting close.

Had Galen been able to conduct vivisections, I suspect he would have figured it out. He was clever, but his cleverness was constrained by the tools and concepts at hand. He was like a geologist who has to infer the history of Earth from the evidence of past events— tectonics and sedimentation. It is possible, but difficult. Even the geologists have advantages over Galen. They can look at a volcano and imagine a historic volcano. They can feel an earthquake and imagine ancient earthquakes. They can watch the settling of sediment on the river bottom and imagine millennia of sediment. Galen could not do anything similar. He could not watch the beating heart or even a pump that might inspire his reasoning. Science often proceeds by analogy or metaphor, and there was no physical device yet constructed that worked like a heart.

It has become popular to look on Galen as having gotten things wrong. He did get some things wrong, but every scientist does; our mark is not where we err but instead where we improve upon previous errors. Galen made far more advances than retreats, and even those who criticize Galen for what he missed live in his influence. Galen's discoveries are everywhere. He was the first Western scholar to take the pulse of patients and use it as a measure of their health. He was the first to urge other doctors to cool down patients with fevers and, conversely, warm up patients with chills (or colds). The idea that physically weak people could be made stronger through exercise came from Galen. Then there were true innovations so radical they would not be practiced again for nearly two thousand years. Galen used a needle to remove cataracts; similar surgeries would not be done again for another eighteen hundred years. He appears to have even conducted brain surgeries in which he removed tumors through holes he drilled in the skull, the precursor to modern versions of tumor excisions.[10] Galen has a strong grip on modern life, just as Roman cities and architecture have a strong grip on modern cities and their design. Galen knew that he didn't comprehend everything, so he kept exploring, trying to understand the body. The quest to prod the body a little more each generation to reveal new truths is his legacy too.

*   *   *

Unfortunately, Rome began its decline just after Galen's death and descended fully into chaos in its Western realms with the death of Romulus Augustus around AD 476. The fire of learning that had been passed, culture to culture, to the Romans from the earlier Greeks, to whom it had been passed by the Egyptians, to whom it had been passed by the Mesopotamians, was extinguished. The light went out. The colonizing hordes were interested only in God or their own satisfaction, depending on the horde under discussion. The papyrus scrolls in the library in Alexandria (including the original Ebers Papyrus) were burned, and with them, the quest to understand the human body turned to ash, demons, and deities.

So began a time that used to be called the Dark Ages, a time during which religion would be valued over all else. Terrible things would happen. Small states would come to war, feudally, against each other. Writing itself disappeared in some places. The medical texts of the Greeks and Romans that survived the fires were neglected. Had all of this happened over the course of one generation, someone might have remembered the old ways. But it was not one generation; the ignorance persisted for hundreds of years, during which knowledge continued to be lost until the point when all of Europe seemed to be inhabited by men and women who knew no more about their own bodies than had hunter-gatherers at the dawn of humanity. They lived in a world of spirits and ignorance where the heart, once again, beat not with blood but with magic. They looked up at the bright moon and saw God. They looked at the sun and saw God. They looked at each animal and plant and saw God, and when they looked at people dying, they figured they saw God too, working the strings of their bodies, snapping what could be snapped, pulling what could be pulled, all beyond explanation or need for corporeal concern. A dark fatalism killed nearly all that had been learned.

Perhaps this is too harsh a depiction. Among historians, it has become unpopular to describe the period after the fall of Rome to around the year 1000 as the Dark Ages. Historians will point out that here and there, pockets of learning continued. Documents were preserved and cherished. Individual small flames were passed, hand to hand, generation to generation—treasured bits of knowledge. This, of course, is what one hopes: that despite all that was being lost on the large scale of Europe, there were individuals who still cherished knowledge. How could there not be someone who wanted to know more? Yet, when it comes to the heart, the opaqueness of the Dark Ages was nearly complete. In the years between 400 and 1400, little new understanding was gained about the workings of the heart, arteries, veins, and blood. Knowledge during this time deteriorated rather than improved. Realistically, less was known about the heart from 1000 to 1400 than was known in AD 400; less was known then, for that matter, than was known at the end of Galen's career, two hundred years after the birth of Christ.

Initially, Galen's work seemed to have been lost to science altogether. In Western Europe, not a single copy of one of his scrolls seems to have survived. But in the eastern part of the Roman Empire, his writings had, it would turn out, continued to be copied and translated, from Latin to Arabic and then from one Arabic copy to the next. Muslim scientists prevented ancient knowledge from being lost in its entirety, not just that of Galen but also more generally. Not all of Galen's millions of words were translated, and meaning and context could be lost in translation, but his flame was passed. When scholars in Western Europe, particularly in Italy, rediscovered these translations, they cherished them—too much. Galen's words seemed so advanced, relative to the knowledge of the time in Europe, that they were treated as literal scripture, wisdom handed down from an ancient that was to be revealed, not built upon. The sciences of anatomy and human biology came to circle Galen, Galen the great, Galen the perfect, Galen the prince.

3

When Art Reinvented Science

A good painter has two chief objects to paint: man and the intention of his soul. The former is easy, the latter hard.

— LEONARDO DA VINCI (CIRCA 1490)

Late one afternoon in 1508, Leonardo da Vinci (1452–1519) was at the Santa Maria Nuova Hospital in Florence, a Church hospital. He was not a doctor, but he already knew more about the human body than almost anyone else who had ever lived, more even than Galen. He was talking with a very old man, a centenarian. The man, who is known to history simply as *il vecchio*, the "old one," was kind and garrulous. He had lived a grand life. Da Vinci had just returned from Milan, and he was dressed in fine clothes—maybe his purple cloak, maybe the pink cape; he was worldly and beautiful. Da Vinci bent over *il vecchio*, his fingers touching the old man's onion-paper skin gently. Then, suddenly, the man died. He died as if struck down. Da Vinci held the man with great kindness before pulling out his knives and beginning to dissect his body. He pulled back *il vecchio*'s clothes and cut into the warm flesh. This was a true autopsy, a word that comes from the Greek for "to see for oneself," which is precisely what da Vinci wanted to do.

Today, we seem set on extending life indefinitely. Death is what we push off at all costs, by whatever means. To da Vinci, the goal of medicine was not to prevent death but, after a good life of reasonable length, to make death, in his word, sweet. A sweet death after a good life was the best that one could hope for. Most ways of dying in da Vinci's time were brutal. Smallpox. Infection. Rabies. Malaria's shivering ache. Or worse. But *il vecchio* had died without pain. He had died of an inevitable and natural process of some sort. But what was it about age that could cause so sweet an end? This was a mystery in which every person alive and every person yet to live had a stake.

The artist began with cuts to the chest, but he moved slowly. The man's body was smaller and more delicate than that of the horses and cows he more often handled.[1] Da Vinci would need patience to see each piece well—each finger, toe, vein, bone, and nerve. No cuts would be spared, and after each cut, a drawing would be done, to understand but also, simultaneously, to improve in the skill of depiction of inner truths. Da Vinci did not know what he was looking for; he was exploring. So little was known about the body. There had been minimal progress since Galen. Anything seemed possible. Perhaps the liver exploded. Perhaps the brain turned yellow. A million, a trillion scenarios lurked in the skin and organs, none of them any more likely than any other. Da Vinci pondered what he saw.[2]

Da Vinci's beginnings, like those of many artists of the early Renaissance, were humble. He was born to an unwed mother in a small town in Tuscany.[3] From there, it appears he was sent to live with his father in the town of Vinci, which would many generations later become known for him, as though the town of Vinci were named for Leonardo and not the other way around. In Vinci, Leonardo would later write, he slept outside in a crib. His life was that of a rural boy. The dark sky rose above him at night. And during the day, he watched the birds wake around him in the trees. One day, a kite landed so near him it brushed its tail on his face.

The details of Leonardo's life seem special now, as they would to him later in life, a foreshadowing of greatness, a series of minor omens. But not then. His childhood was marked mostly by its very ordinariness, until he began to share his art. When da Vinci was about fourteen, his father, a notary, was asked by a man of little means to commission the painting of a crest on his shield. Da Vinci's father gave the shield to his son to paint. Leonardo proceeded to cover the shield with the insignia of a monster, an insignia so marvelous that his father, Ser Piero, promptly sold it for a handsome sum and then fatefully (at least in the context of this book) replaced the shield with another like it, bought from somewhere else, at a lesser cost, emblazoned with a simple heart.

Whether the story of the shield is true or not, at some point, da Vinci's artistic abilities were recognized by his father, so much so that when da Vinci was fourteen or fifteen—and after some arm-twisting by Ser Piero—the Florence-based artist Andrea del Verrocchio (1435–1488) offered the boy an apprenticeship. The apprenticeship was an open door[4] through which da Vinci ran.

The apprenticeship, like nearly everything else in Florence at the time, was a luxury afforded by the city's great wealth. Da Vinci was born at the beginning of the Renaissance. Da Vinci's birth marks the rise of this new age in which knowledge and beauty began, once more, to be cherished. This rebirth is thought of in terms of its art and, later, science, but it was as much a thing of money as of intellect. In Florence, businesspeople, the Medici family in particular, had accumulated enough wealth to allow extravagant expenditures, including the purchases and commissioning of new art. Through their purchases and gifts, the wealthy created a culture in which artists could afford to live by art alone, a culture in which it paid artists to revisit the ancient techniques and even invent new ones. When the ancients were reconsidered, their work seemed perfect, and so it was an obvious first step for the artists of the time to relearn their ways. Once relearned, those ancient ways were, haltingly, built upon; it was out of this building that the art of da Vinci, Michelangelo, Titian, and so many other greats emerged.

Da Vinci trained with his mentor Verrocchio in a workshop in Florence for nearly a decade before he began taking on his own private, signed commissions, work that would pay the bills and more for the rest of his life. The first piece signed by da Vinci comes from a time when he was just starting to work on his own; it is a sketch of the Arno River valley in which water moves through the muscular hills of the valley like, as da Vinci would later note, blood through the heart.

As he began to work on his own, da Vinci benefited from patrons who were both grand in their expectations and, all things considered, remarkably patient. From the beginning, da Vinci worked very slowly. In order to paint, he first had to invent just the right paint. He had to invent new approaches to perspective. And, more than anything, he had to—or at least, he felt compelled to—dissect. Most Renaissance artists viewed dissection as useful training in the details of the body. Da Vinci and others needed to know the body so they could "reveal more effectively its power, fragility and reality in artistic form." Leon Battista Alberti (1404–1472) was expressing a widely held (and essentially Greek) sentiment when he suggested that there were three elements of the body that needed to be studied for painting: the arrangement of the bones, the distribution and arrangement of the muscles, and, finally, the skin and the fat on which it rested. This was also the approach of Verrocchio, Michelangelo (1475–1564), and others. Da Vinci was like other artists of the time in that he studied bones, muscles, and skin, but he was unusual in the extent to which he also examined the rest of the body. He looked into its interstices; each time he performed a dissection, he was, as he would say of himself, like a boy who enters a deep and unknown cave. However frightening exploration could be, he always went further, deeper into the darkness so that he might find his way closer and closer to the truth. He looked to see what was there but also how and to what end it worked. Da Vinci's dedication to understanding the body gave his art a kind of visceral realism others aspired to; it also allowed him to make novel, scientific discoveries.

As he explored, da Vinci made breakthroughs, real scientific breakthroughs, that are acknowledged as such, in a dozen subfields of anatomy—the biology of eyes, neurobiology, reproductive biology, and the study of the blood vessels and heart. Even the very earliest of his surviving studies of anatomy (from about 1485, fifteen years after the drawing of the Arno River valley), a rough sketch on which he would build his future discoveries, displays connections among blood vessels and aspects of organs that had never before been documented.[5] In his studies, da Vinci, like any scientist, could not always be sure what was an advance and what was a rediscovery, but he recorded what he found all the same. Arguably, his greatest achievements were his notes on the heart and blood vessels.[6] Da Vinci's dissections of hearts and blood vessels started with animals: a horse and then cows and more cows. In these animals, the vessels looked to him like the rivers he had explored as a child and loved to sketch as an adult. They ran through the body carrying blood and a kind of as-yet-unnamed magic. Where did they go? How did they work? What moved the blood? Why? In his earliest dissections, da Vinci struggled. He was so influenced by Galen's teachings—nothing better had been done in a thousand years—that he convinced himself he saw forms in the body that were not there, forms Galen had posited. But with time, da Vinci would get better at trusting his eyes, at seeing what was and, based on what he saw, figuring out for himself how things worked. He was never fully free of Galen's tether, but eventually he (and he alone) began to see new truths rather than merely affirm old ones.[7]

As he considered the flesh of humans and other animals, da Vinci came to see the body as a kind of machine, "a vehicle in which to get around and survive,"[8] a vehicle of pumps, levers, and gears whose functions could be understood. Never before had this modern sentiment been so clearly held.[9] In studying this machine, da Vinci, bit by bit, began to pick at Galen's ideas, refining them where they seemed in conflict with what he saw and sometimes simply overturning them wholesale. Galen imagined the blood flowed from the liver (where he believed it was made) to the heart and then on to the lungs, where it was used up, but da Vinci observed that blood flowed through all of the veins and arteries, not just those linking the liver, heart, and lungs. To him, this seemed obvious. It also became obvious to da Vinci that the heart, not the liver, was the center of the system of blood vessels. The heart beat in the embryo. It beat first, and it was primary, the very essence of being alive. The brain was the seat of the soul (today we might say the seat of consciousness), but the heart was its agent, its muscular vessel. Da Vinci was the first to draw the four chambers of the heart accurately and to observe that the atria and the ventricles must contract in concert. He also noted the unidirectionality of blood through the heart's valves and, ultimately, arteries. Blood went through one way only, not back. These modest advances were the first real anatomical progress in more than a thousand years.

Among da Vinci's scientific discoveries in the blood vessels and chambers of the heart, two stand out, the way the *Mona Lisa* and *The Last Supper* stand out among his paintings. One was based on what we would now call a physiological model. Da Vinci looked to rivers and streams to understand the dynamics of flow; flow preoccupied him for the better part of a decade, beginning in 1498. Da Vinci dropped weighted floats, leaves, corks, paper, seeds, and even ink in water-filled tubes in the Arno River in an attempt to grasp how and why water moved. He saw eddies, little circles of water, that often formed when water pushed against the edges of rivers (or, in his experiments, glass tubes). He then drew or painted how these objects moved in order to understand the underlying movement of water.

Da Vinci was intrigued by how blood moved through the heart's valves, valves that, like rocks in a river, could allow or, depending on their position, stop flow. Beginning in 1513, da Vinci made detailed dissections of the heart's valves and nearby vessels while working at a hospital within the Vatican's walls (where he had a modest feud with Michelangelo, who was also dissecting there at the time). Da Vinci's work in the Vatican hospital is what allowed him to see the heart's valves so clearly and to draw them in such great detail. They would not be drawn in more detail until the late 1800s.

Once da Vinci had characterized the physical attributes of the heart's valves, he sought to understand how they worked. He couldn't see the valves functioning in living animals, so instead, he returned to a physical system, the river, to understand a biophysical system, that of vessels, blood, and valves, and to make predictions about its function. This approach of using one system to model another is among the most common modern practices in science, but that was not the case in the time of da Vinci. Based on his river studies, da Vinci predicted, correctly, that blood would move more rapidly through narrower blood vessels. On average, it does.[10] He also predicted that when the big left ventricle in the heart contracted, it would be difficult to keep blood from flowing back through the aortic valve into that ventricle before the valve shut. This valve opens and shuts about once a second. As a consequence, it must seal both tightly and rapidly. Da Vinci thought the answer to how this occurred was in vortices, little eddies of blood that formed in part because of bulges in the aorta just above each valve (later to be called the sinus of Valsalva, for the Italian anatomist). Da Vinci tested his prediction by blowing a glass version of an artery in which he could watch the movement of grass seeds in liquid; it was, in essence, an artificial aorta. In watching the seeds, he saw his ideas confirmed, at least to his own satisfaction. He imagined that as blood flowed through the valve at the exit of the left ventricle, eddies of blood would form to help to close the valve. He

49

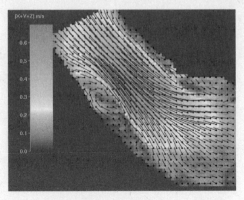

At left, a simplified version of the vortices based on da Vinci's depiction. *(Courtesy of Jennifer Landin)*. At right, a computer model of the same, produced some four hundred years after da Vinci's death. *(Copyright © 2014. Courtesy of Tal Geva, MD)*

was right, though no one would realize it until 1968, when two engineers, Brian and Francis Bellhouse, confirmed da Vinci's prediction using a method essentially identical to the one he had used: they built an artificial aorta and watched the movement of artificial blood.[11] When they published their article, the Bellhouses assumed they were the first to notice these vortices. It was only a year later that they discovered that da Vinci had beat them to the punch by four hundred years.[12]

Da Vinci's big revelation, though, came from what he saw in *il vecchio*, among the best-documented dissections of an entire human since ancient Alexandria.[13] Da Vinci spent so much time with *il vecchio* that the man's body began to rot, stink, and fall to pieces. The work was tedious and unpleasant. As da Vinci wrote in his notebooks (in a bit of extended, if complex, self-praise),

And if you have love for such things [as dissection] you will perhaps be prevented by your stomach, and if this does not prevent you, you may perhaps be prevented by the fear of passing the night in company with bodies flayed and fearful to look upon. And if this does not prevent you, perhaps you will lack good draughtmanship that should belong to such drawing, and if you have the draughtmanship it may not be accompanied by perspective. And if it is so accompanied you may lack the principles of geometrical demonstration and the principles for the calculation of the forces and power of the muscles; or perhaps you may lack patience, so that you will not be diligent.

Da Vinci lacked none of these things, and so in the old man he made many discoveries—even the cause of his death, a cause he ascertained by, once again, availing himself of his river model. By this point, da Vinci had already observed that as rivers grew older, they became more tortuous. A young river traveled down the land in a straight line, but as a river aged, it twisted as it deposited sand along shores. The water did not compress, so where it went through a narrower constriction, it had to move faster, with more pressure. The river would turn and narrow enough that, given sufficient time, it would get longer and longer until the pressure to go straight through a narrow curve was too much and the river would break. In the old man's tortuous arteries, which had turned, twisted, and narrowed with age, da Vinci saw sections of arteries, the body's most important rivers, that were so narrow blood could barely flow through. The man's arteries were more fragile, twisted, and narrowed than da Vinci had observed in the vessels of "birds of the air and the beasts of the field." Those twisted narrowings prevented the nourishment of the blood from reaching the places in the body where it needed to go, and so the body starved. Da Vinci did not know that the blood carried oxygen from the lungs and sugar from the liver. He did not know that without both of these things, the brain would starve in three minutes and the body would die. But he understood the basic consequences of the hardening and thickening of the walls of the arteries. In age, da Vinci discovered, arteries harden and narrow and, as a result, eventually clog. *Il vecchio*'s sweet death was due to what we would now call atherosclerosis.

Da Vinci could have written all of this up—together with equally groundbreaking work on other organs, birth, the biology of the fetus, the articulation of bones, and more—as the most comprehensive book of anatomy since Galen. He discussed the possibility of writing a "treatise on anatomy." In the winter of 1510, he worked with an anatomist, Marcantonio della Torre (1481–1512), perhaps with the goal of producing such a treatise. One may actually have been produced. In 1518, da Vinci showed a document to the cardinal of Aragon. Could it have been da Vinci's great opus, an opus lost to history? Some suspect it was. The only description that remains of the document is what the cardinal's secretary noted:

> This gentleman has written of anatomy with such detail showing by illustrations the limbs, muscles, nerves, veins, ligaments, intestines, and whatever else there is to discuss in the bodies of men and women, in a way that has never yet been done by anyone else. All this we have seen with our own eyes. He has also written of other matters, which he has set down in an infinite number of volumes all in the vulgar tongue, which if they should be published will be profitable and very enjoyable.

Da Vinci appears to have found a strong supporter in the cardinal, a supporter not for his art but, amazingly, for his science. Da Vinci might have gone home to finish the document he showed the cardinal, but unknown to the cardinal, da Vinci had already suffered a stroke. If it was like many strokes, a piece of plaque, secondary to atherosclerosis, was freed by chance movements of blood and the pressure of his beating heart and sent, perhaps along with clotting blood, into the narrow vessels of his brain. There, it wreaked havoc. Da Vinci could no longer move his hand. He would write and draw no more. He died a year later (possibly of another stroke). Whatever book existed was never finished, his knowledge lost except for what remained among his notes.

After da Vinci's death, one of his pupils, Francesco Melzi, began to try to produce from da Vinci's writings a version of the long desired treatise on anatomy. Melzi worked his entire life on the project, attempting to simply restate and convey what da Vinci had discovered. But it was too great a task. Da Vinci wrote backward on the page, having taught himself to write, and Melzi had to teach himself to read backward. But this was not the only problem. Da Vinci also made up words, fused words into unusual compounds, broke ordinary words into pieces for no apparent reason, and never used punctuation. (All of this suggests that whatever the cardinal read was something more polished than da Vinci's notes.) But there was actually an even bigger problem. In reading da Vinci now (in translation and from left to right), we understand his insights, but to his contemporaries, the most interesting of his discoveries were too radical. It has been speculated that Melzi struggled to distinguish the bits of da Vinci's work that seemed to be the spouting of a madman from those that were genuine discoveries. Whereas da Vinci had pushed the envelope of science with confidence without, apparently, caring what his colleagues thought, Melzi cared. He did not want his mentor to seem ridiculous. He knew that much of what da Vinci wrote was genius, but was it all? He couldn't tell. As a result, he never finished. When Melzi died, in 1568, da Vinci's notes (at least the five thousand sheets Melzi had) were still in disarray. Melzi had compiled more than nine hundred chapters based on the notes, but they were a mess, a failed attempt to understand the man who had discovered more than anyone before. There was never to be a great da Vinci book of anatomy. Instead, da Vinci's contributions to science, so close to transforming human understanding, were simply and progressively lost, not to be seen again until after his biggest discoveries were made anew by other scientists.

Surely, da Vinci discussed his insights with his colleagues.[14] He must have shown off some of his drawings. One does not dissect an old man without at least mentioning it in conversation to friends over wine, which da Vinci purchased in huge quantities, according to the grocery bills of his that historians have compiled. *You would not believe his arteries*, we might imagine him saying, but we can only imagine. In the decades after da Vinci's death, his notes disappeared both physically and from discussion. Some may have been destroyed. Others were turned over to colleagues such as Melzi, to wealthy families, and to religious leaders. Some of these notes would be rediscovered and translated in the late 1800s. Others waited until the 1960s. Many of da Vinci's contributions may still be missing. As a result, da Vinci's greatest insights—about the flow of blood, atherosclerosis, and the working of the heart's valves— were hidden and so seem to have played little or even no role in the rebirth of knowledge about the heart. Da Vinci was, ultimately, "a man who awoke too early in the dark."[15] As the scholar Kenneth Keele wrote, "His gigantic efforts in the realm of what we now call science tragically failed to disturb his fellows from their slumbers." If he had any influence at all on the science that followed him, it was likely due to his art rather than his careful dissections. In the *Mona Lisa*,[16] *The Last Supper*, and every other painting by da Vinci, art patrons find greatness in the consequence of brushstrokes. It is a greatness of paint built on the scaffolding of da Vinci's understanding of flesh and bones and even the vital beating heart.

4

Blood's Orbit

ndreas Vesalius was just nineteen years old when he left Belgium. It was 1533, and he believed himself destined for greatness. He considered, very briefly, where he might go to best attain this greatness, and without a doubt, there was but one place: Padua, Italy. In Padua, science was being reborn. There has never been a more important place in the history of science, particularly medicine, than Padua and probably never a more important time than the decades in which Vesalius lived there. Vesalius would go to Padua to be great.

Padua is located in what is now Northern Italy, but when Vesalius arrived, it was a small city in the kingdom of Venice. In Venice, the doge—the leader of the republic—was a kind of god-king in a silk gown and triangular hat. In the late 1400s and early 1500s, a succession of doges, along with wealthy community leaders, sought to make Venice a scientific power by funding education. Venice was already a maritime and military power. Venice was mighty, and soon it would indeed be scholarly too. By the mid-1500s, after da Vinci's death, eager young philosophers—anatomists included— began to flock to the city. As they did, Padua's reputation spread.

Vesalius imagined he would find this flourishing greatness in Padua, and in some fields, it was indeed present, but not, from Vesalius's perspective, in anatomy. In Padua, anatomy theaters seemed to be more like circuses than temples of knowledge. Dogs crawled beneath bodies; everything stank of rot. But what was worse, at least to Vesalius, was the response of the students and professors. Vesalius expected them to look eagerly at the bodies before them, greedy for new understanding. Instead, they looked away. In the dark basement of the Hospital Dieu, for example, Vesalius watched as a professor read from a book while one assistant dissected the body and another held aloft or pointed to the relevant parts. Nothing was being illuminated, not the body's features and definitely not the truth. This was twenty years after the death of da Vinci, and yet nearly every one of his lessons seemed to have been lost.

The science Vesalius found was not real science; it was a kind of education masquerading as science in which anatomists demonstrated the knowledge acquired centuries before rather than attempting to reveal new truths. This form of educational anatomy had its formal origins in 1315. A woman from Bologna was sentenced to death and executed; a professor, Mondino de Liuzzi, retrieved her corpse and, with the Vatican's blessing, began to cut into it in order to teach his students about the body. When earlier dissections had been performed, they were done with the goal of autopsy. In one episode in 1286, for instance, a disease of some sort struck down hundreds of chickens and people in Cremona, Parma, and neighboring towns. In order to ascertain the cause, many chickens and several humans were dissected. What was found was described as a sort of boil or bump on the heart, which is hard to interpret in terms of modern medicine. This was enough to lead to a local pronouncement that no one should eat chickens or eggs,[1] but it, like other autopsies, produced little in the way of formal education or new science. With Liuzzi's dissection, however, informal procedures for dissections began to be established. Each would last four days, after which point the smell of the body would simply be too awful for anyone to continue. Dissections would be held once or twice a year using the bodies of criminals, such as the woman from Bologna, and the body was to be fully dissected. (This differed from previous dissections, which tended to attempt to preserve the form of the body so that the family might still visit the corpse after the dissection.) On the basis of these educational events, Liuzzi wrote *Anathomia*, the first real book about the existing knowledge of dissection and anatomy. The goal of Liuzzi's book, just as for the dissections themselves, was not science as we know it today; it was, more simply, to reveal and educate. His book was a textbook, and to the extent that advances were made, they were advances in how to perform a dissection, not in the biology of the body. The dissection when done perfectly did not reveal truths. It revealed God. Later, when cuts into the heart uncovered only muscle where God might have been expected, the dissections revealed the knowledge of Galen. Even this revelation was imperfect. Liuzzi's book contained a less accurate depiction of the anatomy of the heart, for example, than did Galen's work more than a thousand years before. This was the tradition of academic dissections; it was a tradition far different from that in which da Vinci thrived, a tradition focused on rediscovering the past in each of its details, however flawed.

At first, Vesalius obeyed the norms of this culture, of looking and learning, but eventually, he became too frustrated. He wanted to discover, and he knew that discovery was far from over. More must lurk in the organs than had been revealed. He began to do his own dissections, apart from Liuzzi's traditional two a year. In doing so, he was, at least by his own accounting, the lone learned man actually studying the gore of the flesh. His colleagues discouraged him, and he was threatened (physical fights were not uncommon among anatomists). He bit the hands that trained him. He accused his mentors of "never having lifted a knife except to cut a steak." He cut one body and then another, and as he did, he tried to understand what others had missed, about the heart, about everything, building up personal knowledge from cadavers, wresting discovery from decay.

Vesalius had to cut into a large number of bodies both to see what was present and to understand whether what he had seen was normal, the state of the average body rather than some exception.

But he also had to *find* bodies. Vesalius became an expert at body snatching,[2] among history's greats. He would go to the gallows himself and cut the ropes on which a man hung, letting the weight of the corpse fall against his body. He would go to cemeteries filled with the half-buried bodies of the dead. He forged keys to break into ossuaries. He and his students knew no limits in their search for the dead, a search that broke so many rules that Vesalius would, on some occasions, need to instruct his students to flay the skin of a body before beginning a dissection so that it would not be recognized by relatives searching for the missing remains of their loved one.

As he cut, Vesalius began to doubt more and more of the wisdom his mentors had handed down to him. Some of these concerns he came up with on his own. Others might have trickled down through other sources, even possibly from da Vinci. Despite their never having been published, da Vinci's observations might still have had some influence, perhaps by word of mouth or even in the form of an unofficial manuscript of some sort. Writing fifty years after the death of Leonardo da Vinci, Giorgio Vasari (the great and sometimes accurate biographer of the Renaissance artists) noted that Vesalius "was wonderfully aided by the talent and labor of Leonardo, who made a book drawn with red chalk and annotated with the pen of the subjects which he dissected." We can't be sure what exactly these red chalk drawings included, or that they existed, or that Vesalius saw them, and yet the possibility of a connection between the two men cannot be ruled out. We do know that if Vesalius did see these drawings, they could not have been very complete, since most of what da Vinci discovered would ultimately be missed by Vesalius. Vesalius also made some discoveries that da Vinci missed.

One of Galen's certainties that bothered Vesalius was the claim that blood moved through pores from the right to the left side of the heart. Vesalius cut hearts regularly. He did not see pores; he saw only the thick muscular wall separating the left and right ventricles. Da Vinci had never commented on the pores. To Vesalius, if such pores existed, they were invisible. Or else Galen was wrong. It was still taboo for anyone to openly discuss the problems in Galen's texts. But Vesalius had had enough. Emboldened by frustration, Vesalius wrote the sort of book da Vinci should have written — now considered the most important book in the history of medicine.[3]

In this book *De Humani Corporis Fabrica* (*On the Fabric of the Human Body*), Vesalius revealed the body anew. In describing the book (published in 1543), scholars often note that Vesalius corrected more than two hundred errors in Galen. This is true — sort of. The book was an act of defiance, and it corrected much that was wrong in Galen, but in many cases, it did not do so through the *words* of Vesalius. The revolution Vesalius led was subtler.

When Vesalius wrote his big book, he had no trouble finding an artist willing and able to depict the body. The artist's identity continues to be debated, but evidence suggests it may have been the great Titian himself or, if not Titian, an artist from his school. Titian (1488–1576) was the most renowned painter in Venice in the early 1500s. He was, relative to his contemporaries, "the sun amidst small stars," and he shone for both his use of color and his depictions of human bodies in movement. (Note: His sun shines on. One of his paintings was recently purchased for seventy-one million dollars.)

Whoever the artist was, he drew upon the techniques of Venice and Florence and, especially, da Vinci, including his use of red chalk. Here was, perhaps, a creeping but woefully modest influence of da Vinci by way of his impact on the art of Florence. Vesalius had Titian or someone from his school, someone with a hint of da Vinci's style, depict the body as he, the artist, saw it, although tempered by Vesalius's interpretation. It was in these depictions that the body's realities most clearly deviated from Galen's teachings. Vesalius did not have to say what was wrong with Galen's work if the artist could show it. For Vesalius, as Arturo Castiglioni put it, anatomy "evolves with drawing, for drawing and through drawing. The need of texts was no longer felt so much as the need of pictures,"[4] radical pictures.

Even this subtle rebellion against tradition was not well considered. Vesalius's own teacher Sylvius said of him that he was "an impious madman who is poisoning the air of Europe with his vapourings." To that comment and the many others like it, Vesalius grumbled, "Professors think it beneath their dignity to take a knife in hand." But all that was beginning to change. Vesalius had stormed the hill. It was just a matter of time before a few others tumbled over, though perhaps more time than Vesalius might have guessed.

Why was Vesalius, after fourteen hundred years, able to see (some of) the errors in Galen when everyone else had missed them for more than a thousand years and for more than fifty years after da Vinci? Partly, it was the bodies themselves: having them and having so many of them. It was also his prepared mind; he was blessed with unusual powers of observation. Others were beginning to see partially what he was seeing most clearly. But art too was key. Art was the tool with which Vesalius could see and show both the heart and, in great detail, the arteries and veins, including the very first drawings of the valves in veins. Once Vesalius had shown the heart, more discoveries became inevitable. Vesalius revealed Galen's knowledge to be incomplete, and when he did, other anatomists looked to *Fabrica* and wondered what else had been missed.

The next advance after Vesalius was that of one of his students, Hieronymus Fabricius. Fabricius accurately noted that there were small valves throughout the veins. To Fabricius, the valves seemed to block the flow of blood out through the veins toward the extremities. He was wrong. He, like all those before him, did not understand that blood goes only one way through the veins: to the right side of the heart (the exception is the pulmonary vein, which takes oxygenated blood from the lungs to the left side of the heart). Because he didn't know which way the blood moved, he could not quite make sense of the valves. Yet the observation of the existence of the valves was to prove very important.

The work of Vesalius and then Fabricius combined with the wealth of Padua attracted even more scholars from around the world. They brought with them new ideas and perspectives, and they took home with them new insights. It was to Padua, Fabricius, and the operating theater in Padua called the Bo, in which dissections had become grand public performances, that a young man named William Harvey would eventually come for the education that would help him make what has been called the single largest discovery in the history of the human body. There, the young Harvey would hear from great minds like Fabricius as he looked down upon stinking bodies, their parts emerging and disappearing in the flicker of the candlelight that bathed the theaterlike room. It would be enough light, though just barely, to see big, holistic truths.

William Harvey was born in 1578 (some thirty-five years after the publication of Vesalius's *De Humani Corporis Fabrica*) to a working-class family in Kent, England, where his chances of achieving intellectual greatness seemed slim. He was not a noble, and only nobles had access to the best schools. But precisely at the time that Harvey was readying himself for school, things began to change. For reasons of politics and, again, economics, the educational system had started to open up, which allowed young William to go to a better school than he might have otherwise—a school that, if he succeeded there, could allow him to go anywhere, even college.

Early on in his education, Harvey did not seem marked for greatness. He was not the smartest boy in his classes. But he had one advantage over his peers. While they were brought up in affluence and given what they asked for (including their education—a fancy degree was no guarantee of any sort of real academic accomplishment), Harvey had to work for his successes. Just as his father worked hard in day labor, Harvey would work hard at his education. It was not an avid striving so much as it was young Harvey's understanding of what one should do. One must work, and so he

DR WILLIAM HARVEY.

William Harvey, the man who revealed how blood circulates through the body and, in doing so, reordered our understanding of each human organ. *(Courtesy of the National Library of Medicine)*

worked. He worked hard at school and studied, and when he was done studying, he studied some more.

The first opportunity that came to Harvey as a result of his hard work was medical school. He graduated in classics, rhetoric, and philosophy from Gonville and Caius College at Cambridge in 1597, and he could have continued his studies there, but he got word of an opportunity in Padua, at the Bo. Padua offered him the chance to be among the greatest minds of the time, men (and a few clearly remarkable but poorly chronicled women) who gathered to display their knowledge. In the Bo, Fabricius dissected bodies while, in a room in the same building, Galileo wrote about and dissected the skies.

When Harvey arrived in Padua, he found an environment in which progress was more possible than anywhere else in Europe. But where Vesalius drew some influence in Padua from painters, Harvey's influence would come from another direction, a new generation of astronomers who believed in making predictions and testing them with observations (and even experiments). Galileo was in Padua as a professor while Harvey was there as a student. It is possible Harvey may have taken classes from Galileo, but even if Galileo never directly mentored him, Harvey was influenced by the older man's approach to science. Galileo achieved many firsts in astronomy, but among the most significant was that he built on Copernicus's belief that the earth moved around the sun. Copernicus theorized that the earth circled the sun, but Galileo took Copernicus's work further by generating and then testing a series of hypotheses of the phenomena that would be expected were the sun central. Harvey would come to do something similar for human bodies. He would develop ideas about how they might work and then test, with measurements, observations, and even experiments, the predictions of those ideas.

Harvey thrived in Padua, complaining far less than had Vesalius before him, and he graduated in 1602, having seen by that time, many hearts, dead hearts, hearts as still lifes that he could depict but not quite yet understand. From Padua, he initially took a position as an assistant physician at St. Bartholomew's Hospital in London. With time, he would be given a lifetime appointment as the Lumleian Lecturer at the Royal College of Physicians. In this post, he was allowed a relatively leisurely schedule. He lectured twice a week on anatomy and surgery. This cushy position afforded him the affluence of time, the same sort of time da Vinci required for the flowering of his own greatness. Harvey had time enough to do his job, healing people, but also to explore ideas too grand to be justifiable to any reasonable (or perhaps even rational) employer—ideas about the heart.

With time, Harvey began to make the sorts of observations of the heart and blood vessels that he thought would lead him to new ideas and, eventually, to tests of those ideas. He used an approach we would now call comparative biology. He looked at the bodies of many animals to understand their differences but also to acknowledge that certain phenomena were easier to study in some organisms than in others. Harvey's comparative approach allowed him to see phenomena others had missed. He then combined his observations of these phenomena with the findings of his contemporaries and the ideas of his forebears.

By the time Harvey returned to England in the early 1600s, at least one scholar had offered a wild new hypothesis about the workings of the heart. Yet while the hypothesis itself was stimulating to Harvey, the story of that scholar's life might well have summoned a different emotion—trepidation at the possible consequences of having ideas at odds with the status quo. In Spain, Miguel Serveto (born about 1511), a philosopher-anatomist, had been able to break free of Galen and, in doing so, see the body anew. Between forays into questions about the meaning of life and religion, Serveto discovered a phenomenon everyone else had missed, namely, that the heart pumped blood into the lungs or, as he put it,

The vital spirit is generated by the mixing of the inspired air with blood, which goes from the right to the left ventricle. This blood transfer does not occur through the ventricular septum, as usually believed, but through a long conduit crossing the lungs. The blood is refined and brightened by the lungs, goes from the pulmonary artery to the pulmonary veins, is mixed with the inspired air and eliminates residual fumes. Eventually the whole mixture is sucked by the left ventricle during diastole.

Serveto's discovery rested on several perceptive observations. He noted that the blood entering the lungs was different in color than that leaving it (due, we know now, to the absence or presence of oxygen). He also noted that the artery leading to the lungs was wide and thick, as though a lot of blood flowed to the lungs, not narrow, as might be expected if it was simply used by the body to supply some form of nutrition to the lungs.

Serveto would have been an excellent correspondent for Harvey, a living inspiration. The two might have exchanged insights and built on Serveto's discovery. But such a relationship was not to be. Decades before the two might have had a chance to meet, Serveto died—of unnatural causes. In addition to studying the heart, Serveto had also been engaged in trying to reform religion (Serveto's anatomical insights were actually published in a book that focused primarily on religion); he attacked aspects of Catholicism and Protestantism alike, revisiting the standard interpretations of Christianity just as he revisited the standard understanding of the heart. He wrote a series of books, beginning with *Errors in the Trinity* and culminating in an opus in which he directly attacked Calvin and the idea of predestination (that is, the concept that one's fate is chosen by God before or at birth). Serveto boldly, some might say too boldly, sent this latter book to Calvin. In response, Calvin wrote to Serveto, "I neither hate you nor despise you; nor do I wish to persecute you; but I would be as hard as iron when I behold you insulting sound doctrine with so great audacity." Serveto responded to Calvin, and the correspondence continued until Calvin had had enough. He wrote to his friend William Farel on February 13, 1546, "Serveto has just sent me a long volume of his ravings. If I consent he will come here, but...if he comes here, if my authority is worth anything, I will never permit him to depart alive." Serveto did indeed travel to see Calvin, and on that trip, he was imprisoned by supporters of Calvin. Then, on October 27, 1553, at the age of forty-two, nine months after having published his work on anatomy, he was burned at the stake, surrounded by his heretical writings, including his hypotheses about the workings of the human lungs, arteries, veins, and heart.[5]

When Harvey discussed the heart with students and did dissections, he mentioned Serveto, but initially as an example of errant thinking rather than insightful progress. Serveto's ideas were still too radical. But the more Harvey studied hearts, the more he began to think Serveto might have been right. And if the blood indeed flowed into the lungs rather than passing through the wall of the heart, Harvey began to wonder what other bits of the understanding of the heart would have to change. A Paduan anatomist, Realdo Colombo (1510–1559), had written that the valves of the heart seemed to allow blood to move out of the ventricles only (not back in as well, as had long been supposed), implying that there could be just one direction that blood moved in the heart: into atria, out ventricles. Colombo also showed that the veins leaving the lungs contained blood (rather than air), a reality that had begun to be suspected but was not yet well described. With time, Harvey started to teach his students about the ideas of Serveto and Colombo; he also showed them other evidence that seemed to him to suggest Serveto and Colombo were right. In the veins of the legs, there were valves. Vesalius thought the valves kept blood from pooling in the feet, but if the idea that blood flowed out of the heart was right, the function of the valves could be to prevent blood from flowing backward. So Harvey did a test of the kind Galen had done, a test of the function of the valves and of the direction of the flow of blood. He applied a tourniquet to the arm of a volunteer just tight enough to block the flow of blood through the veins (the arteries, being deeper down, were not affected). When he did, the section of arm below the tourniquet (toward the hand) swelled, as would be expected if blood was able to enter that part of the arm but could not leave it. Harvey also showed that if the tourniquet was applied very tightly—tight enough to stop blood flow through both veins and arteries—blood did not build up in the veins (because no blood could get in through the arteries). Finally, he showed that if an artery alone was blocked, blood would build up above the tourniquet, on the side of the artery closest to the heart and the body's core. It looked as though blood was moving toward the heart through the veins,[6] and away from the heart through the arteries. Harvey just didn't understand how.

Piecing the observations of Serveto, Colombo, and others together, Harvey started to rethink what circulation through the whole body might be like. He imagined new models that might make sense of what he was seeing. As he did, he repeatedly subjected the predictions from these models to real-world challenges, updating the story or shoring up his own confidence with each new test. This approach seems obvious today—creating an idea of how the body might work and then testing it through experiments and observations—but for generations, it had not been used (apart from in the work of da Vinci).

One of the phenomena Harvey tested was how the muscles in the arteries are constructed and work. Galen (at least, as interpreted by his translators) imagined that the flow of blood through the blood vessels was caused not by the contraction of the heart but by muscular contraction of the arteries. Galen and others had felt the arteries pulsing in their patients. The arteries do indeed pulse, but when Harvey studied them in detail, he noted they were reinforced from the inside, not the outside (and blood spurted out from them when they were larger, not smaller). If the arteries squeezed to push blood around the body, then one would expect they would be surrounded by a sheath of muscle. Instead, the muscle—or at least a layer of tough and fibrous cells—was on the *inside*, as though to protect the vessel against pressure from somewhere else. That somewhere else, Harvey thought, must be the heart. Harvey went on to show not only that the heart beat, but that it did so in two steps, first the atria, then the ventricles. Harvey was able to see these contractions by studying the hearts of fish and frogs, in which the contractions were slow enough to observe.

If the heart pushed the blood through the lungs and then back out to the body, a new question arose: Where did all the blood come from? This was a question that even the combined insights of Servetus and Colombo could not answer. Galen thought that after food was digested in the gut, the result—a sort of magical energy—was transported to the liver, where it was converted into blood. In this model, a person needed to eat food in quantities sufficient to equal the flow of blood through the heart. But there was another possibility that Harvey had begun to contemplate when considering the arteries and veins, namely, that the blood was cycling through the body, being invigorated by the lungs and then used again and again. The difficulty of Harvey's proposition was that it required blood to move from arteries to veins somehow, somewhere. But the vessels did not connect, not visibly anyway. Harvey believed it had to happen via some invisible pore—not like Galen's invisible pores in the heart, but maybe another set of invisible pores throughout the body. He could not prove the existence of such pores, but he could, he thought, show the impossibility of their absence by estimating the amount of blood the body would require if blood wasn't being recycled. Da Vinci had calculated the amount of blood, at minimum, that moved with each tick of the pulse. He then calculated the number of times the pulse beat in a day (not distinguishing yet whether it was the pulse of the heart or the arteries). The two could be multiplied to get a crude lower estimate of the amount of blood a liver would need to produce: thousands of liters of blood a day. The estimate was astonishing (the modern estimate of the volume of blood that goes through the heart each day is even greater, six hundred thousand liters). It was a great pouring of liters of blood, too much for the liver to produce and too much for the average diet to supply food for.[7]

Pulling all of this together in a book, *On the Motion of the Heart and Blood in Animals* (1628), Harvey argued that the blood circulated: it was pushed by the heart, went out to the body through the arteries, returned to the heart through the veins, and was pumped to the lungs, where it extracted some vital force; then, with that force, it was pumped back out to the body. He was, of course, right.

Today's textbooks often emphasize that Harvey discovered circulation, but he did more than that. Until one knew that the heart pumped blood from the lungs out to the body and then back and that blood contained vital stuff, there could be no realistic explanation for the function of the liver, the kidneys, or, really, any other organ of the body. Once Harvey determined what the heart, arteries, and veins did, the rest of the body could finally begin to make sense.

The lungs, for example, the organ for exchange of gas with the environment, suddenly had a job. We know they allow the veins to let go of a toxin (carbon dioxide) and pick up a vital element (oxygen). At the most basic level, the lungs give us invisible, vital succor and let out an invisible and deadly waste.[8] As Harvey put it in the paragraph in which he rehung the pieces of the solar system of the body's organs:

> Since all things, both argument and ocular demonstration, show that the blood passes through the lungs and heart by the action of the auricles and ventricles, and is sent for distribution to all the parts of the body, where it makes its way into the veins and pores of the flesh, and then flows by the veins from the circumference on every side to the center, from the lesser to the greater veins, and is by them finally discharged into the vena cava and right auricle of the heart, and this in such a quantity or in such a flux and reflux thither by the arteries, hither by the veins, as cannot possibly be supplied by the ingesta, and in much greater than can be required for mere purposes of nutrition; it is absolutely necessary to conclude that the blood is in a state of ceaseless motion; that this is the act or function which

the heart performs by means of its pulse; and that is the sole and only end of the motion and contraction of the heart.

Much of Harvey's work had been done on dogs, but it was now obvious that within each dog and each human, the heart's rivers flowed in a circle. Galileo had shown without a doubt that the earth circled the sun, and Harvey had now shown that the blood circled the body. This same circuit can be found in every living bird and mammal species and, with some tweaks, in every reptile, amphibian, and fish too.

One might imagine that, with this knowledge of the heart's circuit, physicians would begin a great wave of new research, perhaps even attempt surgeries on the heart. None of this occurred. After Harvey offered his revelation, he was lauded. He retired, and then, so it seems, the entire field followed suit. Harvey's colleagues tried to get him to come back to work, but he stayed in retirement. As he put it,

> Would you be the man who should recommend me to quit the peaceful haven where I now pass my life and launch again upon the faithless sea? You know full well what a storm my former lucubrations raised. Much better is it oftentimes to grow wise at home and in private, than by publishing what you have amassed with infinite labour, to stir up tempests that may rob you of peace and quiet for the rest of your days.

Harvey had begun to be believed, but he was tired. He would leave the next steps to others, and there were many steps. Blood itself would not be understood for decades. Harvey never really understood blood. He never saw how it cycled in the body, and he could not see the capillaries that connected the arteries and the veins. It would take another generation before the Italian scientist Marcello Malpighi would use a microscope to observe the connections, a single cell wide, between the narrowest arteries and the narrowest veins. Like Harvey, Malpighi chose to work on those organisms in which he might most easily see the phenomenon he was interested in. He looked at model organisms. When it came to seeing the smallest arteries and veins, frogs were ideal, and it was in them that he observed with his microscope the capillaries that connect the smallest arteries, the arterioles, to the smallest veins, the venules. The walls of these capillaries are a single cell thick—each tunnel just one four-thousandths of an inch wide—but they are everywhere. No cell in your body is more than twenty microns (about a third of a hair) away from a capillary. Capillaries abut the alveoli of the lungs and pick up oxygen and release carbon dioxide. Capillaries bathe the cells in the body's sea of blood.

But what did the blood actually carry? Harvey ducked this question and simply moved the magic of the heart to the blood. The blood carried vital spirit, inhaled in the lungs and spread to the body. Harvey thought that the heart required this vital stuff to live. He also imagined that some sort of fermentation occurred in the veins just before the blood coming back from the body reached the heart. Fermentation, he knew, turned grape juice to wine. Perhaps it also turned blood that had been used by the body back into something more vital.

I'll pause to note the enormous loveliness of the idea that our blood is enlivened by fermentation. Of course, it is wrong. The character of the blood that reenters the heart is determined not by fermentation but by what has happened to the blood throughout the body as it has been carried here and there. The oxygen and sugar are used up by the cells. In their place, the blood receives cellular waste and carbon dioxide (itself a kind of waste). But Harvey was on the right track; although this process is not technically due to fermentation, there were microbes involved.

No less than 3.8 billion years before Harvey suggested that fermentation enlivened the blood, life evolved on Earth.[9] We take it for granted now that our cells require oxygen, but this reality was discovered only in the mid-1900s and it might easily have been otherwise. When life began, not only was oxygen not necessary, it scarcely existed at all. Although just how life began is the subject of active, albeit speculative, research, we know that life almost certainly began with a single cell. From that cell, every living thing that has ever lived descends. There are no exceptions on Earth, not even the hint that one might plausibly look for exceptions. That first cell and its early descendants, as near as anyone has been able to determine, could deal with extreme heat and lived off the energy they could derive from chemically transforming inorganic molecules such as hydrogen and sulfur. At least initially, their diet did not include other organisms, for the simple reason that there were none. There was also not yet oxygen, and so they were anaerobic (*an-* means "without" and *aerobic* "using oxygen"). They did without, as do many kinds of single-celled organisms living in modern low-oxygen environments, from the muck at the bottoms of swamps to the partially digested food in your colon.

A next major transition in life was the evolution of eaters, species that ate other species. These new life-forms had enzymes—basically chemical knives—that allowed them to digest other cells. Although each was just a single cell in size, they were predators. Suddenly the world had become rougher. Human bodies still use the same genes for many enzymes that these creatures evolved, including those that break down some carbohydrates. Our ability to eat other species, in other words, is a direct consequence of the ability of those ancient single-celled organisms to do the same. Then came the change that set the stage for the origin of hearts.

Three and a half billion years ago, some cells evolved the ability to eat the sun: they photosynthesized.[10] The chemistry of the big bang had hamstrung Earth's denizens, creating a situation in which they had to live off what was left over, the detritus of the universe, but now that species could derive energy from the sun, life itself could add to what was on Earth. Life could even change the atmosphere, albeit without intention or plan. By 2.8 billion years ago, photosynthesizing cells had divided and divided and filled the sea and shallow swamps. They had become so abundant—a green slime of success—that the atmosphere fundamentally changed. It was now full of oxygen, the oxygen that all photosynthetic organisms, be they trees or bacteria, produce as waste when they make sugars out of light's energy, water, and carbon dioxide. Oxygen was toxic to many species, but microbes can make do, and they did. Some lineages of bacteria evolved the ability to use oxygen in their metabolisms; they did so through a process called respiration. These oxygen-dependent species were the most efficient species that had yet lived (though they could still be damaged by oxygen, just as our modern cells can). They thrived, as did the predators that evolved to eat them. Then something really very unusual happened, a one-of-a-kind event.

For the earliest predators, predation occurred as it does now. Predators found, ate, and digested their prey. But at least once, it did not happen quite that way. A predator consumed an oxygen-loving organism, a single individual of a single species, and that individual, like Jonah inside the belly of the whale, survived. More than that, when the predator reproduced (by dividing), the cell inside it reproduced too, so effectively that each of the descendants of the predator had an oxygen-eater inside it. The predators with the oxygen-eaters inside them, one might imagine, would do worse than those without. But they did not; they did better. They were able to produce more energy from the food they digested because they could use oxygen. As a result, these unusual hybrids thrived.

The success of these hybrids has continued for billions of years. There is one of these oxygen-eaters, in fact, inside every human cell. We now call these oxygen-eaters mitochondria; they are the powerhouses inside cells. All eukaryotes on Earth—including plants, fungi, protists, animals, and many more basic kinds—descend from these chimeras; it was a chimerism that begat success in no small part because it allowed organisms to be more active, to get more energy from one piece of food.

As single-celled organisms evolved into multicellular organisms and, eventually, into all of the multicellular organisms we know today, many things changed. Plants evolved all of the particularities that make them plants (they gained their ability to eat the sun from another ancient coming-together of two lineages, one in which a small photosynthetic bacterium was engulfed but not digested by another organism). Sharks evolved to be sharks, and humans, humans. But through all of this, a mitochondrion—that ancient oxygen-eater—stayed in each cell. New challenges arose as organisms got bigger. In single-celled organisms, oxygen diffuses into the cell from the environment. There is no need for more elaborate contrivances to get oxygen. But as organisms got bigger, oxygen did not diffuse well into those cells farthest from the surface. Branching vessels called tracheae evolved to allow oxygen into the depths of newly evolving bodies. Initially, these organisms had no hearts or contractions of any kind, and they were not filled with blood or even its primitive antecedent. It was enough for the tracheae to make a space through which oxygen could arrive and carbon dioxide could leave.

But there were limits related to the levels of oxygen in the environment. For reasons not yet fully understood, eventually the levels of oxygen in the atmosphere declined. When there was less oxygen, organisms needed to go to greater lengths to find it, and some body types were simply no longer possible. When dragonflies first evolved, for example, the concentration of oxygen in the air was as high as it has ever been. Breathing was easy. The dragonflies, which lack a way of pumping oxygen to their mitochondria, could evolve to enormous sizes—some over two feet wing to wing. But when oxygen levels declined, this type of dragonfly disappeared, never to be seen again. It needed, it would turn out, a heart.

Vessels filled not with air but a liquid—like blood—would be a key first step to getting oxygen around a body. For a body much larger than, say, a roach, blood vessels alone are insufficient to get oxygen to each and every mitochondrion. It was in this context that the heart evolved, first as a sleeve of squeezing cells, then as a small pump, then as a pump with two chambers, and eventually with all of the variety we see among animals. The pump worked to squeeze the oxygen in the blood to each place it needed to go. The lungs evolved to provide greater surface area across which oxygen could be absorbed. And then, when even that was not enough, the body evolved the ability to muscularly inhale and exhale, pulling the universe in with each breath. Once the lungs evolved, blood could carry things other than oxygen and carbon dioxide. It could carry compounds, called hormones, that could signal from one part of the body to another. Blood also came to help stabilize the conditions of cells. A bacterium floating in the ocean is at the mercy of the environment, buffeted about by the universe. When it is cold in the ocean, bacteria are cold. The blood of the body buffers its cells (each of which, remember, is in essence a bacterium), making sure oxygen is constant and regulating, at least in birds and mammals, body temperature.

It is amazing that from a chimera of two kinds of microbes the complexity of our bodies evolved, but it is also a challenge. It is a challenge because every one of our cells needs oxygen. It is a challenge because anything that interrupts the cycling of blood interrupts everything. If blood stops cycling, oxygen stops reaching the cells. Because the brain requires the most oxygen—it has the most mitochondria—it is the first to feel the effects. It grows starved. The cells die. Da Vinci did not know the history of microbes or oxygen. He just understood that if the body was starved of blood, it would die. Blood is moved to each cell by the push of the heart until the heart stops, be it sweetly, as da Vinci would wish, or otherwise.

The mitochondria and their history and role were not really discovered until the 1970s. Their discovery meant that the human body was really multitudes held together by the glue between cells and by the heart and its circulatory system, a system whose real goal is to keep alive the great community of cells that each of us represents. This multitude includes our cells, the mitochondria inside them, our partner bacteria inside our guts, the bacteria on our skin, and much more. Death is always the result, in one way or another, when the circulatory system fails to provide oxygen, food, and messages to this diversity of cells. Death is the body's falling apart, its failure to stay connected. We see this most obviously in hemorrhagic strokes. Hemorrhagic strokes are caused when blood vessels in the brain are blocked or rupture. A rupture occurs when a blood clot travels into the narrow vessels of the brain and causes a dam, on one side of which pressure builds up. The pressure leads the vessel to rupture. The rupture, in turn, causes blood to flow directly over the neurons of the brain. This, in and of itself, can cause some damage, but the real damage is that anything downstream of where the vessel ruptures fails to get oxygen and dies. Whatever is contained in those cells—whatever mix of memory and function—dies too. With a big stroke, whole parts of the brain are lost (and with a microstroke, just pieces). This, in the end, is what happened to da Vinci. It is also what happened to Harvey. I suppose one could find some poetry in the fact that Harvey's demise related to the failure of his circulation to reach every part of his body. But what one more obviously sees is the sadness we all face when the heart fails to reach every cell. While Harvey lived an extraordinary life of discovery, he died an ordinary death.

At the time of Harvey's death, no one could do anything to treat problems of the circulatory system or problems related to the heart. Problems happened, and an individual died. Ultimately, there were many reasons for this. But the first one was that no one could detect a problem in the heart or elsewhere in the arteries or veins until it was too late. Strokes destroyed great minds. Heart attacks struck men and women dead where they stood, sat, ran, or danced. Only then did anyone know anything was wrong. Progress in dealing with the heart would have to wait for technology to catch up; it would have to wait for a way of seeing inside a living body and for someone wild enough to try.

5

Seeing the Thing That Eats the Heart

A man sees in the world what he carries in his heart.[1]
— JOHANN WOLFGANG VON GOETHE, *FAUST*

The tragedies of life are largely arterial.
— WILLIAM OSLER

One evening in 1929, Werner Forssmann sat at his favorite bar staring into the distance. From all appearances, he was more forearm than frontal lobe. He was a big man, a football lineman of a fellow. But he had an idea, and as the bubbles of alcohol removed his inhibitions, he decided to announce it. As his friends leaned in, he told them plainly: I am Werner Forssmann. I am twenty-five years old and I am going to insert a tube into the arm of a man and run it all the way to his heart. When I do it, I'm going to fix what is broken in him. "It will," he roared, "change the future of the heart; I will change the future of the heart."

He went on to explain what he would do. His friends were both enthralled and disconcerted. What they were hearing was unbelievable. Forssmann told his rapt and inebriated compatriots he had seen drawings in which veterinarians had run a catheter through the jugular of a horse, winding it all the way to its heart to detect the horse's heartbeats.[2] He found the drawings in an old textbook he had happened upon in one of his cupboards.[3] When Forssmann saw the drawings of the horse experiment, he imagined a whole new science and medicine of the human heart in which the heart could be studied and treated while still beating without anyone having to open the patient's chest. He clung to the sketch. Like a toddler with a favorite doll, he took it everywhere.

At the time Forssmann was contemplating this procedure, there were great barriers to its success. The heart was still viewed as both fragile and inviolable; it was the sacred and untouchable chalice of the body, much as it had been four hundred years prior in Italy. The horse heart could be studied, but it was enormous and sturdy (and, in any case, no one had dared to follow up on the earlier horse work). Not so the human heart. Also, Forssmann was a physician but not really a surgeon. He had obtained his medical degree just a year before. He applied to be a surgical resident at the Moabit Hospital in Berlin but had failed to get the position. He was then hired as a house officer in the Department of Gynecology at the much less prestigious

Werner Forssmann, while doing research as part of his doctoral dissertation in 1928, pauses to look out the window, cigar protruding proudly from his mouth. (*Courtesy of The Werner Forssmann Family Archives*)

Auguste-Viktoria Hospital in Eberswalde, Germany (his getting even that job appears to have taken a favor called in by his mother, who was friends with the chief surgeon). There, he slowly worked his way into the surgery department, more as an assistant than as a surgeon. At best, he would have been considered an unofficial resident. At worst, he was a custodian with a medical degree who was occasionally humored with a chance to use a knife. He did not have a lab or an office. He did not even have keys to the building. What he had, though, he had in spades: thuggish determination. He was ready to be the bull in the china shop of the heart. Horns down, he charged.

Early in his career Forssmann spent many hours around corpses, part of being the low man on the totem pole. He performed rudimentary autopsies. He cleaned up the mess of dying. At first, the bodies disgusted him. It was a natural reaction. But with time, the disgust wore off, replaced by a tingling awe. So recently alive, the bodies were often intact in every way except for the beat of their hearts. In looking at dead bodies, Forssmann saw what others were beginning to highlight: that the hearts of many of the bodies were in bad shape. He could push his fingers into their valves. When he did, the valves felt hard. Often, the coronary arteries—those medusoid snakes on top of the heart—were nearly blocked by a hard white substance that had accumulated inside the walls of the vessels. Whether or not their hearts killed them, the bodies had been ready to go, each one sustained by only a narrow passage through which blood was forced to pass.[4] Without knowing it, they had been walking tightropes with their eyes closed. At the time, and as would be the case for another thirty years, heart problems were viewed as fatal, a fall from the highest rope with no net. One could only watch. The heart stopped. The woman fell. The audience of doctors stood by and looked on. But it was worse than the tightrope. One could detect heart problems only in the dead. In the living, they were both invisible and untreatable. The fall from the tightrope went unnoticed until the body smacked the ground.

But Forssmann, as he pushed his fingers into the cadavers, began to think that some of the problems of the bodies were simply due to bad plumbing that might be fixed. The trouble was how to do the fix, how even to find what was wrong in the first place. Heart diseases were for the coroner to note rather than for the doctor to cure. Forssmann wanted to change that. The same procedure used on the horses might allow him to see some of what was going wrong in living humans, and, just maybe, mend it by releasing medicine into the heart or manipulating the heart muscle itself. How different could the hearts of horses and men be from each other?

Forssmann wanted to make progress, to push ahead. But he was also a conflicted man, a man caught repeatedly between dreams and bosses. As his daughter would later write of him, he was "a daring man who struggled with his passions, an adventurer who searched for pure reason, a tragic man who believed it was his duty to do what he thought was right to do regardless of the consequences."

Forssmann imagined sending a device up through the arm vein of a patient, over the shoulder, and into the heart (the neck vein was also a possibility for entry, but Forssmann thought patients would object to the potential of scarring on their necks, where he would have to cut, vampire-like, into their veins). He would use the veins rather than the arteries because he could move the catheter in the direction of the blood; the catheter would pass with the blood through the valves that Vesalius had observed in the veins on its way to the heart. It would, he blindly hoped, pass without being stopped by the body's ancient security doors.

No one thought the proposal was remotely reasonable. Across history, most radical thinkers confronted with a moment like this, when everyone doubted them, have backed off. Forssmann knew only to charge; it was belligerence more than bravery, a knuckle-headed leap. In 1929, after much arguing, Forssmann finally convinced his boss, Dr. Richard Schneider, of the value of his ideas. He showed him the drawing of the horse. His boss agreed to let

Forssmann carry out the experiment, but only on animals—rabbits.[5] In response to Forssmann's pleas to try it first on a human, on himself, Schneider replied, "Drop this suicidal idea. What could I tell your mother if one day we should find you dead?"[6] A scandal would ruin Forssmann's family. Schneider was friends with Forssmann's mother, and even if Forssmann did not worry about himself, Schneider was worried about her fate, as though Forssmann were a rambunctious teenager acting out.

Forssmann did not want to carry out the experiment on animals; he wanted to do it on human patients. He could have just waited, but he was not that kind of man, and so it was that he decided to do the first experiment on himself. He just wanted to see if he could successfully get a catheter to his heart, like the veterinarians had done in the horse. The rest would follow. The idea was to insert the catheter and then take an x-ray to see how far the catheter had gone. As far as anyone knew, such a procedure stood a high chance of proving fatal. The heart was fragile, its caves delicate and easily disturbed—the glass palace at which one should not throw stones. Forssmann seems not to have cared much for his own safety, but he could not do the procedure entirely on his own. He had to enlist at least one other person, since Forssmann was not highly enough regarded to be given keys to the operating-room cabinets.

At the time that Forssmann was working, very little innovation in the understanding of the heart and cardiovascular system had occurred since da Vinci or Harvey, all those hundreds of years prior, although there had been a little progress in medicine. Perhaps the greatest leap was that scientists now had the ability to see the heart and blood vessels in living bodies. The discovery of x-rays allowed the shape of the heart to be seen in living patients. In long exposures of x-rays on film, the heart showed up as a white body in outline behind the striped cage of the ribs. But that was not all. Dyes of different sorts, if allowed to course through the blood vessels, blocked x-rays. In doing so, and when the images were shown in positive instead of negative mode, they produced black shapes every place blood or other substances could course. The more blood, the darker the image. This method allowed beautiful and haunting images to be created of the brain and its main arteries, of the hands, and of the legs and their twisted rivers.[7] But these methods work best in the dead, among whom diagnosis is often, well, a little late. They were too dangerous in the living.

Forssmann imagined that if he could get to the heart, he could apply useful drugs, but he could also release small doses of dye, doses small enough that they would not kill the patient and could be used to see the patient's heart, after which its problems might become obvious, and some of them might even be mended. When asked about his idea, Forssmann discussed the possibility of using his method to apply drugs to the heart. At the time, the only way to get drugs into the heart was to blindly stab at it with a needle.[8] But the truth was, he wanted to explore. He imagined that there was more he could do in the hearts of patients, if he, if anyone, could only see.

Forssmann stumbled forward another step. It was a drunk's walk. He just needed to convince someone else to help, which he promptly did. He found an eager OR nurse, Gerda Ditzen, and slowly wooed her, convincing her of the beauty and importance of his idea and how it would change humanity. He started to, in his words,[9] "prowl around her like a sweet-toothed cat around the cream jug." He showed her the drawing (he showed everyone the drawing). He explained the value of the procedure to the future of medicine. He even made the outlandish suggestion that he could perform the procedure on her. "Are you absolutely sure," she said in response, "there's no danger?" "Absolutely," replied Forssmann. Ditzen decided to help. She would let Forssmann do the procedure on her. She looked at him and said, "I put myself in your hands."[10]

On the day of the procedure, Ditzen pulled out the equipment—gauze, painkiller, sutures, and a urethral catheter, the piece of tube typically used to drain urine from the bladder. It was going to have to do; he would send that up a vein. Ditzen readied everything. They had not yet tried this method on an animal, and, contrary to Forssmann's later assertions, they had not even tried it on a cadaver. It was consensual madness. Forssmann had her lie back on the table, where he strapped her arms and legs down (for her safety, he said) and numbed her arm. She was ready.

But while Forssmann numbed her arm he simultaneously, without letting her see, moved the surgical cart behind her head, sat down, and numbed his own arm at the crease in his elbow. He was going to do the procedure on himself after all! He paused briefly as he looked at his veins. He looked at the thick rubber tube he was about to insert. He had time to reconsider—he almost did—but this wasn't the moment to stop, and that wasn't his nature. And so he cut into the vein in his left arm, without Ditzen noticing. He then took the tube, with the needle at one end, and snaked it through the incision he had made and up; it was only then that Forssmann learned that there were no pain fibers inside the vein (it was, until that point, entirely possible that the endeavor would be ferociously painful). The tube moved easily, in the direction the blood flowed. It passed through the valves that Vesalius had so long ago detected. He pushed quietly in such a way that, at least initially, Ditzen did not suspect what was happening. She called over to ask when he would start, at which point he told her he was done—mostly anyway. Ditzen saw what was going on and screamed in frustration and then watched as he pushed the catheter farther and farther up his arm. He had pushed it far enough that it had made its way past his elbow and then to his shoulder and then farther still, up and over his shoulder and down toward the right atrium of his heart (he chose the left arm because the vein from the left arm takes a less sharp turn toward the heart than the vein of the right arm), but then he stopped just shy of finishing the job. There was a problem: Forssmann and Ditzen had chosen a room without an x-ray machine.[11] Forssmann could feel his progress but could make no permanent record of it. He had no proof!

Forssmann unstrapped Ditzen and begged her to call an x-ray nurse. She did, and then they began the unthinkable: together, they started walking to the x-ray department in the cellar.

With a piece of metal bobbing near Forssmann's heart, many things could have gone wrong. Forssmann walked alongside Ditzen out the door, toward the stairs. He then walked down two flights of stairs. He reached the bottom of the stairs and walked into the room with the x-ray machine. The second nurse, Eva, was there waiting, as was Forssmann's friend Peter Romeis, who was furious and worried. He tried to rip the catheter out of Forssmann's arm. Forssmann kicked Romeis in the shin to get him to step back (Forssmann's arms were occupied). Forssmann, it seemed to Romeis, was frothing mad.

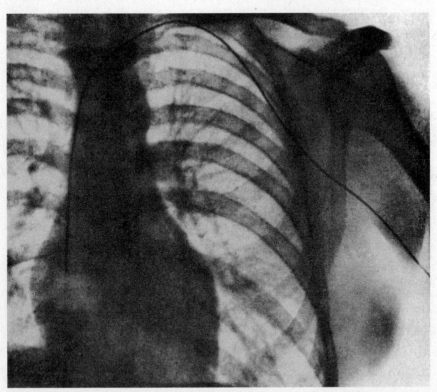

The original chest x-ray of the catheter lead in the right chamber of Forssmann's heart. With this lead, Forssmann touched his own heart. *(Courtesy of The Werner Forssmann Family Archives)*

Eva then took two pictures. But the catheter was still in his shoulder vessel, not quite in the heart yet. She paused, and Forssmann shoved it up a little farther, until it reached the right atrium. Its tip just hung in; he growled with success. Eva took another picture. In it, you can clearly see the lead hanging in his heart, touching its living interior. He had proof. His life, he imagined, was about to change.[12]

Quickly, news of the procedure traveled up and down the halls of the hospital. Forssmann was still high with excitement when his pissed-off boss, Dr. Schneider, called him into his office. Forssmann showed his boss the images, and the boss paused, his rage subsiding. Forssmann had done something amazing. Schneider, too, knew these experiments must continue. Forssmann was opening up an entirely new field of science. That night, Schneider took Forssmann to "an old-fashioned, low-ceiling wine tavern where the waiters wore formal evening dress." They had good food and a lot of good wine and celebrated the birth of a new field of medicine, what would come to be called cardiology.

Forssmann repeated the procedure on himself five more times, and then on a patient, using the catheter to release medicine into her heart. He also practiced on dogs, although, because he could not find any facilities to house the dogs, he kept them at his mother's house, and she kept them in the bathtub (there, they were less likely to soil her carpets). Forssmann would sedate a dog in the bathtub and then drive the dog to the hospital to insert a catheter in a procedure that, just as with his own body, nearly always worked. The method was not just diagnostic; it could be used to cure. Both Forssmann and Schneider were hopeful that the new method would allow doctors and scientists to study and mend living hearts. But Forssmann still needed a real job; he could not advance the science of the heart working as an unofficial resident. He applied this time to Charité university hospital. The chairman of the hospital, Professor Ferdinand Sauerbruch, was the leading surgeon in Germany at the time; he would go on to pioneer a number of techniques for surgery on the heart himself. He was very skeptical of Forssmann's procedure but allowed Forssmann to work at the hospital, without pay. One month later, Forssmann published a paper with Schneider about his new method, announcing it to the world (and boldly lying about the details of what he had done, to make it seem less insane).[13] The article attracted attention in newspapers around Europe. This infuriated Sauerbruch. Forssmann, he thought, was turning medicine into a circus. Sauerbruch fired Forssmann from the job for which he had not even been paid in the first place.

Forssmann never found another job in surgery. There would be moments of hopefulness and opportunity, each of them followed by defeat. While Forssmann's work was very popular with the press, other surgeons, particularly in Germany, thought it too outlandish. Forssmann was exiled from his own discoveries, left to poke around bodies and dream big ideas while spending most of his time helping more senior doctors, conducting exams, and treating patients with ordinary maladies. His bold steps had amounted to nothing. Then World War II began, and Forssmann was sent to work as a medic on the front lines.

What Forssmann did not know when he went to war was that in America in 1940, two physicians, André Cournand and Dickinson Richards, had read about his work and begun to follow up on it, refining his method and turning it into a simple and often-used tool. Cournand and Richards independently figured out how to use Forssmann's technique to reach the left atrium (by puncturing it with a fine needle). They also figured out how to measure blood gas concentrations in both atria. Their approaches shone light on the story of respiration and also began to reveal problems with the major chambers of many people's hearts, problems Forssmann had imagined might be seen, though probably not quite so easily. By the end of World War II, Forssmann's catheter was being used in U.S. hospitals to release dye into the chambers of the heart to visualize, via x-rays, how the heart was working.[14] What began as a barroom brag had turned into modern medicine in the United States just as Forssmann was working hard to save lives on the wrong side of the most horrible war.[15]

Forssmann does not appear to have been aware that his method had moved into common use in the United States. He was too busy surviving. After the war, things had gotten even more difficult (American troops found him at the end of the war pleading for his life, having no possessions other than his copy of Goethe's *Faust*, which he clutched to his chest). Forssmann was held as a prisoner of war until 1946, whereupon he traveled with his wife and six children to a small town in the Black Forest. There, he did whatever he could to survive. Initially, his wife, who was a urologist, supported the family. Forssmann applied for a job at the local hospital but was rejected because he was Prussian. He was refused a loan to start a private practice too. He cut wood, and then, as he always did, he tried again. Finally, in 1950, four years after his release, he found a job as the chair of a small urology department in Bad Kreuznach. At this point, he and his family hoped for nothing more than a peaceful life.

On October 11, 1956, Forssmann was again in a local pub, this one in Bad Kreuznach, when his wife phoned to tell him to return home immediately. Someone with a foreign accent had called. Forssmann didn't care. He went back to drinking. He got home many hours later, at ten. There was another call; it was from Bonn. Someone wanted to interview him. He refused. The next day, he woke up and went back to work. He operated on two patients with kidney disease. He did his daily labor. That day, he heard that two Americans were going to be awarded the Nobel Prize for developing heart catheters, his technique, his telescope into the body. He didn't feel anything. He was still numb from this news when the medical director of his hospital found him in the operating room and quietly announced, "Mr. Forssmann, I would like to be the first to congratulate you and your wife. You have received, with two Americans, this year's Nobel Prize in Physiology or Medicine."

The prize let the world know what he had done. His family left Germany for the first time to receive the award. In a picture taken in Stockholm, a seated Forssmann looks at his wife, who sits beside him in her own chair, her arms folded over her dress. Behind them stand the six children. It is a happy moment. One of the boys is even making a silly face.[16]

Forssmann was finally being given what he had so long felt he deserved. He might have ridden this success into contented retirement, but Forssmann's life was never simple. Upon his return to Germany, Nobel Prize in hand, he applied for new positions. Initially, once more, he had no luck. Then some came. He was asked to head a German cardiovascular research institute and to perform open-heart surgery, the sort of job he had dreamed of his entire life. He declined. It was too late; he did not have the skills the heart

The Forssmann family before the Nobel ceremony in Stockholm. The children, Renate, Bernd, Jorge, Knut, Wolf-Georg, and Klaus, stand behind Forssmann and his wife, Elsbet. (*Courtesy of The Werner Forssmann Family Archives*)

demanded, at least not anymore, and would not develop them. In 1958 he was offered another position, the chair of surgery at the Evangelische Krankenhaus in Düsseldorf, where his role would be more administrative than creative; he accepted. There he worked until his retirement in 1969, at which point Forssmann and his family moved back to the Black Forest village where they had lived after the war. There, the Nobel Prize sat in the living room like a trophy from a high-school football championship. Ten years later, Forssmann died of a heart attack, but the truth was, his heart had been broken for decades. Months before his death he had commented on his life that "it was very painful. I felt that I had planted an apple orchard and other men who had gathered the harvest stood at the wall, laughing at me."[17]

Forssmann brought about a fundamental change in what could be seen in the heart. Whether or not he was able to appreciate them, the fruits of his orchard would nourish the next decades of cardiology. In 1991, after the reunification of Germany, the hospital in which he first performed his procedure was renamed the Werner Forssmann Krankenhaus. The room in which he performed his procedure is still used, and as Forssmann's daughter reported when she visited it years later, it is still a long way from that room to the basement. Forssmann's daughter and his sons are his other legacy. They have gone on to professional success around the world. And then there is the legacy of Forssmann's procedure. Forssmann took the first step of gaining access to the heart. He also took the second step of releasing medicine, and he set the stage for visualizing the atria and ventricles using dye, but he did not see the problem with the heart that actually killed him: he did not see the coronary arteries.

Forssmann's technique came to save many lives by allowing surgeons to see the basic features of the heart before they cut, but it stopped short of being able to show doctors everything they needed to see. Forssmann knew that coronary arteries, even his own, could

The hospital in Eberswalde where Forssmann performed his first surgery. It now bears his name. (*Courtesy of The Werner Forssmann Family Archives*)

clog. Forssmann, among others, imagined that clogged arteries might be mended. But the blockages were invisible until death or until the heart was cut open, such as in the cadavers into which Forssmann had put his fingers. Dye could reveal these arteries, but the problem was that it would do so at the cost of death to the patient—clearly not an option. The dye used in visualizing the heart was safe when diluted in the big chambers of the heart, but in the narrow courses of the arteries, it was too concentrated and, as a result, toxic; even a wild man has limits. Forssmann would not release the dye into his own coronary arteries, and so he would not do it on anyone else's either.

Some thought coronary artery surgery would never happen, that the ability to see the heart had crossed the last safe bridge. In thousands of hearts in hundreds of hospitals, catheters found their way to the edge of the narrow passageways to the coronary arteries and stopped, much like surgeons had stopped at operating on the heart itself in the previous century. Progress would come, but it would take an accident, a stumble into those deeper and perilous caves.

The man who stumbled was the chief of surgery at the Cleveland Clinic in Cleveland, Ohio, Frank Mason Sones. By the point in his life at which his story and that of Forssmann intersect, Sones was already famous in the circles of heart surgeons. He was bold. He was ambitious.[18] He was crass, and he worked all the time, a cigarette hanging out of his mouth as he bent over patients. If Forssmann was the roguish lineman of the heart, Sones was its smash-faced, foul-mouthed quarterback. He was a tough leader who took to wearing stained white T-shirts at the hospital. He barked orders.[19] He lavished praise on success and stomped upon failure. He fought openly with the surgeons around him and nearly always won. People either loved working with him or quit.[20]

Yet, while people could say many things about Sones, and they did, no one ever accused him of being dumb or of missing an opportunity. Very early in his career, he became so good at using angiography, Forssmann's technique, for identifying heart problems that his right-hand man, Don Effler, could treat things no one else could, simply because they had been identified; because they could be seen. Sones described research as being "like kicking down a door,"[21] and he had kicked down his fair share. He shared with Forssmann a ferocious intent to move forward, to make breakthroughs that would save lives.[22] Unlike Forssmann, he was also lucky and powerful. But luck can have its limits. One day, Sones was working in the basement of the Cleveland Clinic in a hole he had carved to stand in; the hole was necessary for the enormous image amplifier Sones had requested so he could see and capture supersize views of the heart. When he stood in the hole, he could look up at the x-rays of patients above him. It was one of many things he did to improve his ability to diagnose disease (he also worked with Philips and Eastman Kodak to increase the power of x-rays). On this day, he was in this cave doing a relatively ordinary procedure. He was using a catheter to inject a small amount of dye into the aorta of a patient's heart to examine what seemed to be a case of valve disease (of the mitral and aortic valves). Dye was, by this time, regularly injected into the aorta and the main chambers of the heart to see both the chambers and the valves. It was, in this alone, invaluable. The one thing Sones was incredibly cautious about, however, was allowing contrast dye into the coronary arteries.

The thought, going back to the work of Forssmann, was that injecting dye into the coronary arteries would cause a deadly arrhythmia called ventricular fibrillation, in which the signals that keep the heart's beat lose their rhythm, and the ventricles lose control of their beating, like a bird with one wing beating at a different time than the other.

On this day, October 30, 1958, the patient was on his back, arms stretched out, looking up at cracks in the ceiling. He was awake. He had to be; it was the only way Sones knew if something was wrong and so was standard operating procedure at the Cleveland Clinic. The patient, like all the others, was the canary in the coal mine of his own body. Sones inserted the catheter, as he had done many times. He ran it up and into the heart. It was, based on the x-ray, in the right place. Then he instructed an assistant to release the contrast dye, 40 to 50 ccs, a large dose but an ordinary one. As the dye began to bleed into the heart, the catheter "jumped." It jumped into the right coronary artery and started to spray dye. Sones gasped, "We've killed him!"

Sones climbed out of the hole and started looking for a scalpel to cut into the patient's chest. The man's heart slowed, its rhythmic pacing diminished from peaks to mere hills. Sones yelled for him to cough. "Cough!" The dye would be physically moved out of the arteries by the coughs; some of it anyway, maybe. Today, Sones might have shocked the patient's chest, but even such simple technology had not yet been invented. If his heart stopped all the way, as it was clearly about to, the only choice would be to cut him open, which would take longer than the three minutes one has before the brain loses too much oxygen and the patient dies. The man was going to die.

And then he didn't.

Slowly, the man's heart started to beat normally again, of its own accord. Sones, who had breathed only enough to yell, *"Cough, cough, cough,"* inhaled and began to smile and yell, this time with joy. Something very, very good had just happened. The man had lived, but the more lasting impact would be that Sones had just unintentionally pioneered the main method that is still used to see all of the heart.

Many surgeons in that situation would have paused to be thankful that the patient had lived and then moved on. Initially, Sones too, in his own telling, "could feel only relief and gratitude." But soon he realized he had seen what was possible; he had seen the future. He could inject dye into the coronary artery and not kill the patient. This patient, yes, had been close to dying, but it would be a matter of tweaking how much dye was used and how. Sones wrote of the moment, "I began to think that this accident might point the way for the development of a technique which was exactly what we had been seeking...If a human could tolerate such a massive injection of contrast [dye] directly into a coronary artery, then it might be possible to accomplish this kind of opacification with smaller amounts of a more dilute contrast. With considerable fear and trepidation we embarked on a program to accomplish this objective."

Sones's initial discovery came of a mistake, but as he moved forward to perfect this discovery, he would make no more. The proof would be in the subsequent attempts, of which there would be many. An hour later, he was planning to repeat the "mistake." A mere two days later, Sones did. It worked again. It worked the next time too. Over the next years, Sones repeated the procedure again and again, inventing a new catheter to make it easier and also relying on progressively faster film and better amplifiers along with direct defibrillators (to restart the heart without cutting open the chest) to make everything safer. By 1962, Sones reported on the use of dye in the coronary arteries in 1,020 patients.[23] By 1967, Sones and his colleagues at the Cleveland Clinic had performed the procedure, which has come to be called a coronary angiogram (literally, a "heart drawing," from *angio-*, "heart," and *-gram*, "drawing or writing"), 8,200 times. Sones's fame grew in light of the coronary angiogram. Teaching others how to see the heart's hidden canals gave him pleasure. Yet Sones could not fix the problems he was seeing in the heart. For as fast as his new method of seeing spread, first within the Cleveland Clinic and then around the world, so too spread the realization of something that had been known but not yet obvious: the arteries that fed the heart seemed to clog nearly inevitably. Maybe this was just, as da Vinci had described it, the sweet death, but Sones had seen many heart attacks, and while some could be considered sweet, clean ends to fruitful lives, they were not all that way. But more important, clogged arteries, sweet or not, were not rare.

By 1816, William Black had discovered by looking at cadavers that the cause of chest pain, angina, seemed to be the hardening and thickening of the coronary arteries (da Vinci noted atherosclerosis but does not appear to have ever focused on atherosclerosis of the coronary arteries in particular). Chest pain was the only real sign (short of death) that these arteries had clogged, and so, to the extent to which the incidence of clogged coronary arteries was considered, it was thought to be synonymous with the incidence of angina. But in his angiograms, Sones saw that it was far more common than anyone had thought; atherosclerosis was evident even in those with no symptoms. In a healthy heart, the vessels—the aorta and the main arteries—would all be black. The angiogram would look like a twisted black flower, with the strong dark lines of the coronary arteries arching over the top. The black was the color of the dye indicating blood. But in many of the thousands of people whose angiograms Sones studied, Sones saw white. The white patches were the places where the dye did not go, where the great and necessary arteries were clogged; the white was the plaque, the thing that eats the heart.

There were two problems moving forward. The first was that, at least as of the time Sones started performing his coronary angiography, there were no treatments for heart disease due to clogged coronary arteries—none. The history of treating heart disease was, until that moment, a history of magic and hope. In 1492, Pope Innocent VIII was suffering from angina. By some accounts, he had three ten-year-old boys brought in. The boys were hooked up to the pope so that he could receive their young vital blood orally. The boys and, ultimately, the pope died.[24] Several centuries later, when Charles II of England was found to be suffering from what seems, in retrospect, to have been either heart disease or the early stages of a brain hemorrhage (the mind's rotten kin to heart disease, as we'll explore later), he was bled. By 1958, the fate of someone afflicted with heart disease was little different—the treatment was bed rest and a glass of wine. When a man or a woman came to the doctor suffering from angina pectoris (*angina* is Greek for "strangling" and *pectoris* for "chest"), Sones could now, finally, see what was causing the strangling; he could even diagnose clogged coronary arteries before the strangling. He was just unable to do a damned thing about it.

In 1958 (the date of Sones's accidental angiogram), surgeries of the heart had become more common, but they were still special cases. Wounds could be stitched up. A few congenital problems could be repaired, a few holes mended, but the vast majority of heart problems were simply irreparable. These were humble times,[25] which makes what happened next all the more amazing.

Over the next ten years, surgeons would move into the heart with what in retrospect one must describe as wild recklessness.

Many even said as much at the time.[26] The next step for Sones would be to try to mend coronary arteries clogged by atherosclerosis or to prevent the atherosclerosis in the first place. But something else was needed before he could do so with frequency and ease: someone had to figure out a better way to work on the heart. The heart, as of the time of Sones's discovery, could still be operated on for only three to six minutes. After that, the lack of oxygen to the brain would kill the patient. Six minutes was about twenty minutes too few for ambitious interventions such as those that might mend the coronary arteries.

Someone needed to figure out a way to allow longer surgeries. The first attempts focused on chilling the body and, in essence, slowing everything down so that six minutes might be turned into a dozen. This worked, but not always and only with incredible difficulty (initially, the patient had to be dunked in ice, though later approaches would focus on just chilling the heart itself). But there was another possibility. Some surgeons thought one might be able to create a kind of machine that would keep oxygen moving through the blood while the heart was opened up. It would be, if only for ten minutes, a replacement for the heart, and ten minutes might be long enough to do something about clogged coronary arteries and maybe a great deal else.

6

The Rhythm Method

One night in 1930 at Massachusetts General Hospital, a teaching hospital of Harvard Medical School, John Gibbon's mentor, Dr. Edward Churchill, called Gibbon to the room of a female patient who was pale, tired, and short of breath. The woman had been through gallbladder surgery two weeks earlier. Churchill moved the patient to the operating room and asked Gibbon and a young technician, Mary "Maly" Hopkinson, to watch her through the night, recording her pulse and respiration every fifteen minutes. The patient appeared to be suffering from an embolism (a blockage of coagulated blood) in the pulmonary artery, the big vessel that connects the right ventricle of the heart to the lungs. The embolism was almost certainly caused by the earlier surgery. The predicament was dire. If the blockage remained, it would prove fatal, choking off too much of the blood flow to the lungs (and ultimately back to the heart and brain). But few surgeons had ever successfully removed an embolism and none ever in the United States. Churchill decided he would try, but only if the patient's blockage became complete and she lost consciousness, in essence fating her to death in the absence of intervention.

Gibbon was tasked with the job of alerting Churchill if the patient's condition deteriorated. He was waiting for the patient to nearly die. All night, he and Hopkinson kept vigil, by turns talking with the patient and watching her sleep. Then, at eight in the morning, the patient lost consciousness. Churchill was called. He ran to the room, cut into the patient's chest, and pushed apart her ribs; he had just three minutes until she would suffer brain death. The operation was done by feel; Churchill could not see through the blood spitting up out of the beating heart, but he found the pulmonary artery and, once there, several clots, which he successfully removed. But it was too late. The patient had gone too long without oxygen reaching her brain and never regained consciousness.

Emergency surgery was a kind of tragic circus, one so dangerous no one should ever want to look, and yet, as a young surgeon, Gibbon had to. He looked at the body and cried. He was haunted by that night. As he later wrote, "During that long...vigil the idea occurred to me that the patient's life might have been saved if some of her cardiorespiratory functions might be taken over by an extracorporeal blood circuit."[1] In response to this experience, he might have decided to return to writing or some other career. Instead, he would focus the next years of his life on trying to build a heart-lung machine that would supply oxygenated blood to a patient's organs when her heart and lungs could not do so, at least for the duration of a surgery (initially, the surgery he imagined was embolism removal, but before long he realized that such a device would also permit a whole new field of heart surgery). He also married the technician, Maly Hopkinson, who would collaborate with him on the dream that from the very beginning she shared. Gibbon was a good man who wanted to do great things with his life, with Maly.[2]

Born on September 29, 1903, John "Jack" Heysham Gibbon was fated to be part of a fifth generation of physicians.[3] His family had spent more than a hundred years mending those who needed it, mending and consoling. In that entire time, nothing the family had done for problems of the heart had changed. One ineffective medicine might be favored over another, but the end result was the same. Patients sick at heart were sent home, urged to rest and take one or another potion. Gibbon was born to the first generation in which all of that might change.

When Gibbon entered Princeton University, he did not want to be a doctor. He studied French literature. He traveled through France with his sister. He was, in his sister's telling, "afire" with "intellectual interests and philosophy."[4] He wanted to be a creative writer or a painter. He graduated at the age of nineteen, a free spirit in the world headed for a bohemian life. But his father told him, "If you don't want to practice you needn't, but you won't write worse for having [your medical degree]." Gibbon either was convinced or acquiesced and so entered medical school at Jefferson College in Philadelphia, graduating at twenty-three in 1927.

It was just three years after leaving Jefferson College that Gibbon found himself standing beside Churchill and Maly Hopkinson looking at the dead patient who would motivate him to build a heart-lung machine. In 1930, few imagined a heart-lung machine would be possible, and so when Gibbon, just twenty-seven, and Maly began to consider how to build such a machine, they were inventing from scratch. The two used a lab on the top floor of the Bulfinch building of Massachusetts General Hospital and gathered parts from any lab that offered them. They tested their ideas and makeshift equipment on stray cats they caught together on the streets of Boston; the lungs and hearts of cats are small, and so it was an easier first challenge to produce a machine that would pump and aerate the small quantity of blood necessary. But even for cats, the challenges were considerable. Somehow the machine had to provide oxygen to the blood without damaging red blood cells; it had to be incredibly gentle and forceful at once.

In 1931, Jack and Maly returned to Philadelphia, where Jack took a job as an assistant surgeon at Pennsylvania Hospital (he was simultaneously a fellow at the University of Pennsylvania School of Medicine). Having landed a more permanent job, Gibbon wanted to work on actually building the machine, but there just wasn't much time, and so the dream was on hold (except at the dinner table, where Jack and Maly talked about it constantly). Gibbon's colleagues appear to have liked him and Maly but thought their scheme unlikely to succeed. Three years into his position, Gibbon decided to move, again, back to work with Churchill at Massachusetts General Hospital, where he was promised space and time to work on a heart-lung machine. Maly was promised a position as Gibbon's technical assistant. Once back at Mass. General, within a year, Gibbon and Maly had a prototype. The Gibbons tried it on some more stray cats. At first, nearly all the cats died, and what was more, they died tragic, awful deaths. Gibbon and Maly were demoralized, but by the end of 1934, still their first year back in Boston, one cat survived, with no ill effects, for twenty minutes. Twenty minutes! In their enthusiasm the two did a jig beside the cat, squealing with a sense of what they had done but also of what was to come; they had done it! Gibbon would later say of that moment, "Nothing in my life has duplicated the ecstasy and joy of that dance with [Maly] around the laboratory."

The Gibbons did not publish their findings (they would wait another three years), but word had spread. On the basis of the successes with cats, the University of Pennsylvania Medical School offered Jack Gibbon a position as a surgical research fellow in the Harrison Research Labs. He took it. Once back home in Philadelphia, he and Maly continued to make progress working with both cats and dogs, though they spent less time in the lab themselves and more time managing a team, the sort of team necessary to actually make such a contraption work. The machine was refined. The cats and, to a lesser extent, dogs were brought off the machine more predictably and in a more or less healthy condition. Jack, Maly, and a growing number of assistants were on the verge of having a machine ready to test on humans. And whereas the two originally imagined the machine would be used primarily for surgeons to work on embolisms, it was increasingly clear to all involved that it would also allow an entire field of other heart surgeries, "impossible" surgeries.[5] The machine itself was, as Gibbon put it, an assemblage of "metal, glass, electric motors, water baths, electrical switches, electromagnets, etc. . . . [that] looked for all the world like some ridiculous Rube Goldberg apparatus." But it worked, at least on relatively small animals with relatively small lungs. Then World War II began.

Jack Gibbon volunteered as a reserve officer despite feeling like he was getting closer and closer to the breakthrough to which he and Maly had devoted their professional lives. He was a man drawn to duty. It was, apparently, the other thing his family did. Gibbon's maternal grandfather (the only grandparent he had known growing up) fought in the Civil War, as did a great-uncle, another John Gibbon, who was a prominent Union commander in the Iron Brigade. His father volunteered for World War I and the Spanish-American War.

Gibbon became chief of surgical services to the 364th Station Hospital in the Pacific theater. There, he improvised out of necessity. This improvisation inside the body allowed him to continue to think about his machine and heart surgery even while he was away from the lab. When he came home to his wife and family four years later, medically discharged due to a herniated disk, he was an even more qualified surgeon than he had been before the war. He was recruited for a more prestigious job as professor and director of surgical research at Jefferson College (which he accepted in January of 1946) and readied to continue work on his great machine.

With the end of the war, the global landscape changed politically, but it also changed in terms of science and medicine. The United States had become *the* power in discovery, and this power made many impossible things, Gibbon's heart-lung machine included, more likely. Whereas before the war he often struggled to find support for the work on his machine, after the war, developing the machine became a major endeavor at Jefferson Medical College Hospital, one with which other doctors were eager to assist. In fact, some had "assisted" in his absence (Clarence Dennis had assembled his own heart-lung machine, based on Gibbon's design, and tried it in the operating theater).

With the help of a growing lab, Gibbon improved the machine he and Maly had been working on. Maly and Gibbon had less to do with the hands-on aspects of this new phase of research; they hired and enlisted new colleagues more expert than they were at the details of what was necessary. Soon the new team had a new prototype that, like the one he designed before the war, would work, but still only mostly. The device used a pump to oxygenate blood that then traveled through a series of tubes, cannulas, and valves. But as the blood moved through the pump, it would clog or, worse, pick up infections. Oxygen bubbles also appeared, which could cause brain embolisms and a death worse than the one the machine was meant to prevent. Gibbon was trying to replace the heart's elaborate subtlety with a crude machine, and the lungs' great networks with something even cruder. The machine resembled the heart and lungs in the same way that a man with feathers glued to his arms resembles a bird.[6]

The team was able to fix the pump by replacing it with a device that would squeeze the tubes of the heart-lung machine the way that muscles around the intestines squeeze (by rolling over the tube and then rolling again), which ushered the blood along. But there was still oxygen getting into the heart and then hiding behind the valves. Dr. Frank Alibritten,[7] whom Gibbon had hired, had an idea. He could make a vent for the air, a kind of chimney. The vent would be stabbed through the muscle of the left ventricle.

Then there was another problem. Then another. More improvisation. Gibbon and his team were getting closer and closer to the finish line—trying his device on humans—but it was such a long race.

What Gibbon had hoped for during the years and years he worked on his machine was medical success; he wanted to be able to treat patients. But those all around him had begun to talk about popular success, media success. Gibbon hated the idea of his work garnering publicity. He believed a doctor should heal in anonymity. Gibbon wanted to solve problems, not discuss them. But he had little choice. Even before Gibbon tried to use the machine on people, the media came to him. As Miller Wayne noted in his book *King of Hearts*, "Few spectacles are as beguiling as a man trying to play God with electricity and steel." It did not help that one of Gibbon's "team members" was Thomas J. Watson, the chairman of IBM (the same man after whom IBM would name its most impressive computer, the computer that won against a human on the game show *Jeopardy!*). Watson[8] provided financial support and allowed his engineers to work with Gibbon to make the device more automatic (and robotlike). The collaboration with Watson had the effect of lending the endeavor an even greater science-fiction mystique than it might have had otherwise. *Life* magazine described the machine as a "robot, a gleaming, stainless steel cabinet as big as a piano."

In 1949, the team members tried the robot on bigger animals: dogs. They were able to stop the heart of each of nine dogs and then replace the function of those dogs' hearts with the machine. In one case, this had worked for forty-six minutes. *Time* magazine announced this success and left little ambiguity about the next step—humans. The same year, the National Heart Institute awarded Gibbon and Jefferson Medical College $26,827[9] to speed up the development of the machine. Gibbon's colleagues and the public believed heart surgery was about to change. But Gibbon and his team were less sure. They were still unhappy with the portion of the machine that oxygenated the blood.[10]

Two years later, the opportunity to use the machine on a human arose when Gibbon was operating on a fifteen-month-old girl, again at Jefferson College. Her heart was not beating properly. With her parents' consent, Gibbon and his team opened her chest and inserted a catheter into the two largest veins entering her heart (veins with blood depleted of oxygen). The catheters were connected by plastic tubing to the giant machine, which would add oxygen to the blood in the tubes by way of two pumps, the first of which pumped the blood to the machine's artificial lungs and the second of which pumped it through. A third pump sent the oxygenated blood back into the body through a tube connected to an artery in the girl's groin.[11] Six assistants controlled the machine itself. But something was wrong. Gibbon cut into what he was sure was the source of her problem, her right atrium, where it was expected that she had an atrial-septal defect, a hole between the two atria of the heart. But he couldn't find the hole. He felt around to no avail, searching and searching. The baby bled to death before he could find and mend what was broken. Only later, during the autopsy, did Gibbon learn that the defect in her heart was on the outside, not the inside. He had been searching in the wrong place. He had not done an angiogram; Jefferson did not have the right equipment or team. Gibbon mourned the baby for the rest of his life. But he continued with his machine.

Later the same year, Gibbon and Alibritten were presented with another chance. In January of 1953, an eighteen-year-old freshman at Wilkes College in Wilkes-Barre, Pennsylvania, Cecelia Bavolek, came into the hospital complaining of shortness of breath and an irregular heartbeat. She was (wrongly) diagnosed with rheumatic heart disease (secondary to a streptococcal infection) and asked to return in two months for a checkup. She reappeared in the hospital on March 29 and her symptoms were much worse — fever; chills; an enlarged, murmuring heart. Forssmann's cardiac-catheterization approach was used to examine her heart, and she was found to have an atrial-septal defect, the same sort of heart problem that the baby had been thought to have, the same hole in the heart. Every time her heart pumped oxygenated blood from her left atrium to her left ventricle, a large volume of that blood went through a hole in the septum, the wall separating the two atria, and back into the right atrium. Gibbon scheduled her for surgery on May 6, 1953. He was surer this time of the diagnosis. He and Alibritten readied their machine for the surgery, priming it with donor blood. Anxious about the surgery, Alibritten had been unable to sleep the night prior. Then the hour came: they wheeled Cecelia next to the heart-lung machine and opened her chest. Cecelia's veins and arteries were connected to the pipes of the machine; its oxygenator began working. Everything proceeded more or less according to plan. But the defects in her heart were more severe than Gibbon had thought they would be: the hole was the size of a half-dollar. What he'd imagined would be a six-minute surgery lasted ten minutes, and then fifteen. More time passed. Gibbon began to doubt himself, and then, what was worse, a technical problem arose. The heart-lung machine clogged. Dr. Bernie Miller, who was assisting on the procedure (and who was by then in charge of the day-to-day activities in Gibbon's lab), had to try to get the machine going again; it had run out of blood thinner. Miller was able to keep it going, partially manually, while Gibbon continued to work. Things were desperate. At twenty minutes, they were fourteen minutes beyond the moment when death would have occurred without a heart-lung machine. His hands fumbled to sew shut the giant hole he had found. Twenty-two minutes. He looked at her face. It still seemed flush with blood. Twenty-four minutes. He stitched madly. The nurses looked on; their faces blanched. Twenty-five minutes. Twenty-five and a half minutes. Was Cecelia's face becoming slightly paler? Twenty-six minutes. Gibbon was done. He quickly took her off the machine, and her heart…restarted! It restarted, as did her lungs. She breathed in deeply and recovered well over the next weeks before going on to a long, successful life.[12] *Time* magazine proclaimed that Gibbon had "made the dream a reality." James Le Fanu described the heart-lung machine as "among the boldest and most successful feats of man's mind."[13] Cecelia would live to the age of sixty-six; the heart-lung machine would live even longer. In ever more novel forms, it would live forever.

Gibbon himself had relatively modest hopes for how the machine would be used. As he said to a *Time* magazine reporter

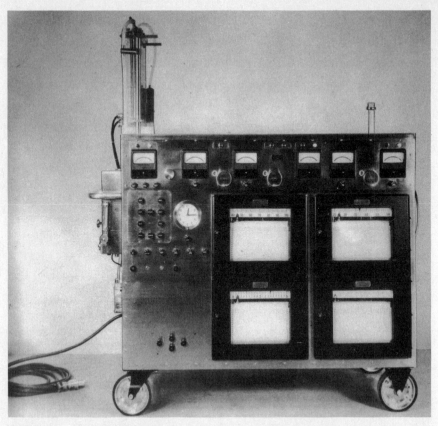

Model II of Jack Gibbon's lifesaving but enormous heart-lung machine, the ancestor of the tiny heart-lung machines found in every hospital today. *(Courtesy of the Thomas Jefferson University Archives and Special Collections, Scott Memorial Library, Philadelphia)*

after the successful surgery on Cecelia, "The machine is not a cure-all for all heart conditions. It will probably be used chiefly on patients born with a deformed heart. It can't help coronary artery disease...But now, for the first time, it is possible to look into the heart."[14] Gibbon was humble in his hopes and humble too in his sense of what he had accomplished. When asked by reporters to pose with Cecelia and his machine, he declined. He was, he said, camera shy.

Gibbon would perform two more surgeries that July using his

Jack Gibbon and Cecelia Bavolek, whose life was saved thanks to the heart-lung machine, one piece of which sits before them, humbly, in the picture. *(Courtesy of the Thomas Jefferson University Archives and Special Collections, Scott Memorial Library, Philadelphia)*

heart-lung machine. His patients, both five-year-old girls, died. In one of the two surgeries, just as in his first, the death was attributable to a misdiagnosis. He grieved for those children deeply; their deaths, he thought, represented his failures, and he never cut into another body. He hired a trained cardiologist at the hospital and built a catheterization lab—the two together might, he thought, have saved the baby and the child who had died due to incorrect diagnoses. But he personally was finished; he never operated on the heart again. According to his friend Willem Kolff, he never even wanted to see the heart-lung machine again. He did not even prominently report the successful operation, burying it instead in a relatively obscure surgical journal. Gibbon eventually came to focus on his private practice and teaching, and then he retired altogether, to paint. Maly, for her part, decided to go back to school to obtain her master's degree in social work (she would later become a marriage counselor). But others picked up the device as he set it down. Gibbon's machine allowed surgeons, for the first time, to work on a heart in which the blood had been drained away and details could be seen. They could work for tens of minutes, and with those minutes, they would try everything possible, and then a few more things.

With a few tweaks (and the benefit of other advances in heart surgery), the heart-lung machine has gone on to allow an entire range of heart surgeries. Through those surgeries, the heart-lung machine is saving lives even as you read. Heart-lung machines are much smaller and almost foolproof, and they can be used to bypass the entire heart (as Gibbon's did) or just the right or the left side. Since the 1960s, most major hospitals have employed perfusionists whose job it is to operate the heart-lung machines. Compared to an actual heart, these machines remain crude, and yet they are profoundly important, a symbol of the power of technology but also of the limits thereof. The device's abilities, after all, are fleeting. The machine is useful for minutes or hours, not days or years. When a patient is on a heart-lung machine, the lungs do not receive blood, and even if the brain survives, the lungs would ultimately die, flutter and fail.[15] Then, of course, when the electricity goes out, the heart-lung machine stops. Even when the electricity stays on, the heart-lung machine can break. Each machine needs tending to, constant repair. A human heart, a real heart, might last a hundred years without maintenance. Yet Gibbon and his team had replicated, however temporarily, two features of the workings of the heart, oxygenation and movement of the blood. With this, he and his team set the stage for many new surgeries that had previously been impossible. But he had also made another step, one toward the replacement, piece by piece, of the functions of the heart.[16]

Gibbon's machine was a temporary electromechanical replacement for one of the cardiovascular system's functions, respiration. But from very early on in the story of the heart, it seemed as though a more permanent electromechanical fix might be possible; it seemed that one might be able to, at the very least, replace faulty wiring in the heart. The heart can short out in many different ways. It is not the only electrical part of the body. Every cell in the body runs on electricity, but the electricity of the heart is special. It is more intense, measurable, and consequential. It is this measurable electricity on which an electrocardiogram (or EKG) relies. *Electro-* derives from the Greek for "amber," which, like electricity, was perceived to have the power to attract. *Gram* is from the Greek word meaning "drawing" or "writing"; think of the word *telegram*. An electrocardiogram represents the telegrams being sent electrically by the heart and measurable anywhere in the body. The first EKGs were performed in primitive form on dogs; the dogs were made to stand in salt water that would conduct (and, ultimately, help record) the electricity of the beating heart. With time, better conductors were developed. Now, a very simple machine can be hooked up to your skin (one electrode on each arm, one on each leg) to record your electrical rhythm. EKG monitors can even be worn on the body, a form of electric fashion. EKGs record the initiation of the contraction of the atria and then the contraction of the ventricles, a predictable hill, then a mountain (with a few small subsequent hills), then a hill again when the ventricles finish contracting. If everything goes right, this topography of the body's electricity repeats for an entire life, billions of rises, billions of falls.[17]

The electricity in the heart triggers the heart to pump in two steps. In the first step, a signal from the sinoatrial node (which sounds complicated but is just a cluster of cells on the top of the right atrium) spreads and simultaneously causes the atria to contract and signals a second node, the atrioventricular, to do the same (which causes the ventricles to contract). The resulting contraction of the heart is like the contraction of a snake swallowing its prey: the muscle fibers encircling each chamber of the heart shorten during contraction and squeeze the blood. The sinoatrial node discharges its signal one hundred times per minute, and so, in order to maintain the slower resting heartbeat characteristic of most human hearts, the nervous system must constantly send signals to slow the heart. When something exciting or dangerous happens, the nervous system just has to take its foot off the brake to get up to a hundred beats per minute (speeding up even more takes more active control).

Each step in this signaling can go wrong. Every day people come into hospitals, have EKGs done, and learn that, indeed, something has gone wrong. Usually, they have a sense of a problem in advance, but not always. The heart can lose its rhythm when its atria, those top chambers of the heart, temporarily lose their beat. Most of us have felt this, especially, common wisdom holds, when having consumed a bit too much coffee. With too much coffee, the atria beat too soon. In doing so, they contract before they have filled and push too little blood into the ventricle.[18] In your chest, this feels like no beat has happened at all. Then, on the next beat, the ventricles are filled with too much blood. You feel this second beat more strongly than normal. The overall effect is "no beat," "too much beat." A related problem occurs when the ventricles contract too soon, before they receive the signal from the sinoatrial node, producing a similar effect: too little blood, then too much. These minor arrhythmias, referred to as ectopy, can be terrifying. Fortunately, they are harmless.

Another form of arrhythmia, atrial fibrillation, occurs in about one in twenty adults over sixty-five. With atrial fibrillation, the atria contract erratically and asynchronously. The muscle fibers in the atria are no longer under the control of the sinoatrial node and contract on their own, like a bag of uncoordinated worms. Without coordinated contraction, blood pools in the recesses in the atria and can clot; these clots can travel to the brain. In addition, because the signals from the sinoatrial node arrive at the atrioventricular node irregularly, the contractions of the ventricles, and hence the pulse, become irregular, sometimes rapid (with associated fatigue, palpitations, or even heart failure), and sometimes slow. The symptoms of atrial fibrillation can be as idiosyncratic as the contraction of the worms of the atria. Atrial fibrillation can happen as a result of aging, due to viral infection, or, most often, for reasons no one understands.[19] In most cases, atrial fibrillation is treated by primitive medicine. The heart is shocked to stop it.[20] When the heart restarts, the hope is that the rhythm will return to normal (no one understands why; this is just medicine's version of rebooting the computer). If this does not fix the problem, part of the heart can be scarred—ablated—with the intent of preventing extra signals from causing one or another area of the heart to misfire.

Just as the atria can become asynchronous, so too can the ventricles. But when the ventricles beat out of rhythm in ventricular fibrillation, the resulting problems are life-threatening. Even if the atria are firing poorly, a fair amount of blood drops down into the ventricles. But if the ventricles are misfiring, the whole heart spasms and fights against itself.[21] With ventricular fibrillation, even slow pumping is not possible. Without the pump of the ventricles, no blood gets to the brain, and without intervention, death is certain.[22] The first step in dealing with ventricular fibrillation is to shock the heart so as to stop the chaos; then, with the heart hope-fully rebooted, the details of the problems can be considered.

In short, the heart's rhythm can flounder in several ways. Its beat is not easy for the body to keep. I describe this complexity because when surgeons started to contemplate the idea of an artifi-cial pacemaker, the number of ways the natural pace of the heart could go wrong was not fully understood. What was understood was that the heart was electrical, and so, with this simple electricity as context, surgeons around the world began, on the eve of Gib-bon's success with the heart-lung machine, to attempt to create an artificial pacer, a pacemaker. The idea was simple: replace the aber-rant natural stimulation of the heart with a more regular, artificial one; this was a solution with, it seemed, the potential to resolve many of the ways in which the electricity of the heart could go wrong. It was a simple fix for a diversity of electrical dilemmas.

One of the men who most eagerly took on this challenge was Wilson Greatbatch. Greatbatch was a tinkerer and an inventor and, at least for a time, a professor of electrical engineering at the Univer-sity of Buffalo in New York. He was no surgeon. He had not even thought very much about the heart until, in 1956, he made a mistake. The same year that Gibbon's heart-lung machine began to be used at hospitals outside Jefferson College, Greatbatch was trying to build a device to record heart sounds for the Chronic Disease Research Institute. As he worked, he reached for a resistor to complete the cir-cuits of the device. Absentmindedly, he installed a circuit of the wrong size, which caused his device to give off intermittent electrical pulses. It was a silly mistake, but rather than just rebuilding the device, Greatbatch paused. He was, he thought, onto something. He suddenly remembered conversations he had had in 1951, while an undergraduate in an animal-behavior lab at Cornell University, con-versations in which researchers talked about the heart's electrical activity and heart block. Maybe, just maybe, his artificial pulse would be enough to trigger the heart to beat. It could do so at any pace one might choose. It could remedy, at least temporarily, the various ways in which the heart's beat could go awry.

By the time Greatbatch made his mistake and grabbed the wrong resistor, doctors and researchers had been shocking hearts for two decades. In the 1930s, Albert Hyman developed a technique to quantify the electricity coming from the heart, and he measured it, accurately, at about one-thousandth of one volt. Hyman reasoned that, in a patient whose heart had stopped, a similar shock might resuscitate it. In this procedure, Hyman shoved a hollow gold-plated needle through a patient's ribs into the right atrium of the heart. He then started his generator. In this way, Hyman saved many patients whose hearts had stopped. He also inspired a variety of tinkerers preceding Greatbatch to contemplate more sophisticated electrical treatments for the heart, perhaps even a long-term artificial pacemaker. If one could jump-start a heart, one might also be able to keep it running.

Greatbatch dedicated himself to making what we now call a pacemaker. He set up shop in a small barn behind his house. He needed to shrink the device down and make it suitable to dwell submerged in the juices of the body. Other pacemakers had been built, but they were all enormous, unreliable, typically connected to car batteries, and, unlike Hyman's gold-plated needle, relied on shocking the entire patient (rather than just the heart). Greatbatch knew he could do better.

Within two years, Greatbatch had a device he thought was ready to try. On May 7, 1958, Greatbatch went to the Veterans Administration Hospital in Buffalo, where Dr. William Chardack, chief of surgery, connected his device to the heart of a dog. The device was called Tiknik 6 (after Sputnik, which had just launched with a dog as a passenger), and to seal it off from the body's liquids, it was wrapped in packing tape and plastic. Tiknik 6 was clunky and did not work long. But the next day, Greatbatch built Tiknik 7, which worked for twenty-four hours. It was only then that Greatbatch realized what others already knew: this was a race. Several other labs in the United States and Sweden were attempting to develop a pacemaker that was small and effective enough to be implanted in a human body. Other teams had already built pacemakers that could be plugged into the wall. Greatbatch quit his job in order to dedicate himself fully to the project. This meant that he and his family would have to live on their savings and the food they could grow in their garden. Greatbatch put aside two thousand dollars for research funds. The family's privations paid off. In 1960, Dr. Chardack implanted Greatbatch's device in ten patients. In every case, it worked.

In October of 1960, Greatbatch sold the device to a fledgling company called Medtronic. Medical electronics companies were not yet flourishing. Their success was a possibility, not a foregone conclusion. Medtronic was founded in 1949 by Earl Bakken and his brother-in-law Palmer Hermundslie. Medtronic started in a pair of garages behind Hermundslie's parents' house in Minneapolis. But things would change. By the end of December 1960, Medtronic had taken orders for fifty of the Greatbatch-Chardack devices at $375 apiece. In 1963, Medtronic had a net loss of $144,135. Sales continued at about that pace for the next few years, and Bakken and Hermundslie talked about selling the company; they still weren't really making money. But they had invested so much time in the company, they couldn't quit (though they could not sustain losses much longer; they were broke). They decided to focus on fewer products and on making those products more perfect; this was an ethos that Greatbatch, the engineer, appreciated. The mission statement became to strive without reserve "for the greatest possible reliability and quality in our products...and to be recognized as a company of dedication, honesty, integrity and service." In the service of this vision, Greatbatch and Chardack continued to work on new innovations to make their pacemaker safer and more effective. Eventually, this approach, as well as Bakken and Greatbatch's personal visits to the client-doctors, paid off. In 1963 the company made $72,923. In 1965, $151,108. In 1969, a million dollars. In 1970, two million. Since then, the company has continued to grow, and with it the number of people whose hearts are paced by machines. In 2012 it had a global revenue of $16.2 billion, a number roughly equal to the GDP of Mongolia, Benin, or Namibia.

With time, Greatbatch's device was joined on the market by a family of others. All of these devices, including Greatbatch's, share a basic working principle. They produce a small, regular electronic stimulus that takes over where the heart's natural pacemaker is failing. How this small shock is delivered depends on the device. In some cases, a wire is run through a vein to the heart. In others, open-heart surgery provides an opportunity to place an electrode and a device directly on the heart. In 2009, more than two hundred thousand pacemakers were implanted in the United States alone. Roughly one in every five hundred adults in the United States has a heart paced artificially by small electrical impulses. What was once unimaginable is now, in many places, ordinary.[23]

Greatbatch never imagined that his pacemaker would become permanent. It seemed to him, from the beginning, to be a stopgap, a temporary fix until the natural pacemaker could somehow be restored. He wasn't a doctor, and so just how that might come about was beyond his reach. Initially, some pacemakers were indeed used as temporary fixes—akin to the heart-lung machine—but it quickly became clear that patients with artificial pacemakers could leave the hospital. They could go about their ordinary lives, with one caveat: the device depended on a battery, a battery that needed to be replaced and recharged.

Greatbatch could not fix the natural pacemaker of the heart, but he thought he might be able to produce longer-lasting batteries; the ones used at the time lasted only two years under the best of circumstances. So he went back to his tinkering barn in Buffalo and got to work. Greatbatch explored a lithium-iodine battery that had been invented by another group of researchers a few years earlier. Immediately, it seemed to offer promise as a long-term solution, with one problem. The lithium-iodine battery had a tendency to explode. Greatbatch tweaked the design again and again and eventually produced a battery that would last far longer than two years and would not explode. Greatbatch founded a company (Greatbatch, Inc.) to produce the batteries. The company succeeded beyond his wildest expectations, and today almost every pacemaker has a lithium-iodine battery.

Greatbatch never stopped inventing things. He pioneered other medical implants, worked on efforts to develop a helium-based fusion reaction to make power, and even invented a solar-powered canoe. In total, he held patents for 325 inventions, and even as he aged, he continued to tinker. At seventy-two, he took a 160-mile trip on the Finger Lakes of New York in his solar-powered canoe. The canoe held him up and powered him along much as his pacemaker did and does for millions of people around the world.

Greatbatch died in 2011 at the age of ninety-two, having extended the lives of people in every country. But even as he approached death, he wished for more, telling an interviewer on the phone, "I'm beginning to think I may not change the world, but I am still trying."

When surgeons looked at the successes of Greatbatch (and, earlier, Gibbon), they came away with the sense that if the oxygenation of the body and the beating of the heart could be replicated, perhaps the same might be true of the entire heart. Eventually, this would yield attempts to create an artificial heart, perhaps one that would even pulsate (heart-lung machines do not). But the heart-lung machine, and an increasing understanding of the ways in which the heart's electricity might be manipulated, also engendered an alternate approach, that of taking the heart out of one person—a heart that already had its own beat and pump, a heart with no need for batteries—and putting it in another, using electricity to bring it back to life or keep the pace. This was not the next step either Gibbon or Greatbatch had had in mind, and yet it was an approach made possible by Gibbon's heart-lung machine, which could keep a body without a heart alive, and a better understanding of what it took to pace a fumbling heart. To most physicians and researchers, the mere idea of heart transplants seemed simultaneously mad and heroic. There were historical precedents for transplants (albeit not of the heart), but they came as much from myth and strange experiments as from modern medicine. The Egyptians, Phoenicians, and Greeks all had myths about creatures composed of half of one species and half of another. Pegasus, for example, was a horse with bird wings, and the Minotaur had the head of a bull and the body of a man. Historically, a Chimera was a mix of a goat, a lion, and a dragon, but in the modern use of the word, these mythical beasts were all chimeras assembled out of pieces.[24] The first suggestion of a more literal transplant occurs in roughly AD 400, when two brothers, Cosmas and Damian, grafted the healthy leg of a very unlucky Ethiopian man onto the body of a man who'd lost his leg to gangrene.[25] In the 1760s, John Hunter, a Scottish surgeon, transplanted a human tooth onto the head of a rooster. He also transplanted the testes of one rooster onto a chicken. Later, in the early 1900s, the French surgeon Alexis Carrel (1873–1944) and the physiologist Charles Guthrie (1880–1963) designed experiments in which they transplanted organs from one animal to another. They transplanted each organ to the outside of the recipient organism's body, leaving the organ to dangle attached from its veins. If any dog needed an extra heart, Carrel could do the job, at least temporarily, giving it one that dangled from its neck vein. Invariably, Carrel's dogs died when the new hearts and other organs were rejected by the recipients' immune systems.[26] One could have taken from Carrel's experiments the lesson that the immune system was a major obstacle to transplants and that it was of supreme importance. Instead, history marked Carrel's experiments as evidence that heart transplants might be possible in humans. Inspired by Carrel, experimenters tried more earnestly, if no less outrageously. The intent of such procedures was sometimes to transplant the attributes of one organism into another. In 1916, John R. Brinkley, a Chicago surgeon and charlatan, began to transplant bits of testes from cadavers into men who wished to be more virile, including himself and an editor at the *Los Angeles Times*. The idea spread and came to include donor tissue from other species as well, including goats, wild pigs, and even deer. Thousands, perhaps tens of thousands, of these transplants were performed.[27] More often, transplants were attempted to replace a damaged body part or organ so that the body, restored piece by piece, might live on and on.

Greatbatch compared his efforts to pace the heart of a dog with the Russians' shooting a dog into space. The men who sought to transplant hearts did not think of Russians floating in space, nor were they particularly interested in dogs. They compared themselves to the American astrophysicists who were, at the time, talking about actually landing a man on the moon, alive.

7

Frankenstein's Monsters

*A thing of immortal make, not human, lion-fronted and
snake behind, a goat in the middle, and snorting out the
breath of the terrible flame of bright fire.*

— HOMER, *THE ILIAD*

Richard Lower's life was one of repeated successes. He had
been trained at Stanford Hospital in San Francisco, where
he met the man who everyone would come to believe had
the best chance of performing a successful human
heart transplant, Norman Shumway. Shumway would become
Lower's mentor, and Lower, for his part, became Shumway's right-
hand man, a surgical resident to his attending surgeon, an assistant
professor to his professor. Together they would pioneer heart
transplants, not as freak-show oddities but as medical realities.

There were many barriers to heart transplants. Some of these
had been overcome before Lower and Shumway began their
work, through the development of angiograms and heart-lung
machines, which allowed broken hearts to be visualized and operated on, respectively. Others were known but not yet circumvented.
But most were not yet even known; they were the Odyssean challenges to emerge during the journey. Shumway and Lower would
travel toward and deal with these barriers using dogs, though that
was not originally their intention.

Lower arrived at Stanford in the fall of 1957. By the summer of

1958, the two men had begun to do experimental surgeries together. They started by trying to develop new ways to keep both the heart and the body alive for as long as possible during surgery. In the fifth-floor lab of the Stanford-Lane Hospital, the two had access to a heart-lung machine and they could use it to operate for long periods on dogs. Dogs have the misfortune to have hearts similar in size to those of humans, and so they were the preferred animal for heart experiments. The pair's first really newsworthy experiment was a kind of endurance test. The two could already clamp off a dog's heart and sustain its body on a heart-lung machine, as could surgeons working on humans. The next step was to keep the heart itself alive. They tried to do this by cooling the heart down to about 28 degrees centigrade, since the cooled heart required less oxygen. And it worked. The heart was alive but disconnected from its dog; the dog alive but disconnected from its heart. Shumway and Lower found they could keep a dog's cold heart alive on ice for ten minutes, and then twenty, and then eventually as long as one hour. This was a breakthrough that seemed as though it would allow any surgery they could imagine, a breakthrough very early in their collaboration, a breakthrough suggestive of the great possibilities that lay before them. The next step was to try for even longer times, but before this happened, the two men got bored.

Some people doodle when they are bored; some eat. Not Shumway and Lower. To take the edge off the tedium, the pair decided to see if they could completely remove a heart from a dog and then put it back in the same dog. In the previous trials, the hearts were clamped off from the bodies but still attached. This was something more, a test of what was possible. As they attempted it, they encountered problems, problems that soon fascinated them both and drew them deeper in. The first problem was that the dog's aorta was very short and its heart was brittle; there was little to sew back together when replacing the heart, and what was there was hard to manipulate.[1] It was a mess. Blood clots formed, and the first twenty dogs on which they tried the procedure died. But then a few lived, and, as they did, Shumway and Lower were emboldened. They were no longer just killing time. The hobby transplants became a goal in and of themselves. At some point it became clear they were working toward doing transplants of a heart from one dog to another. Inasmuch as Lower and Shumway cared about dog hearts only to the extent that they were models for human hearts, they were working toward the transplants of a heart from one human into another.

Shumway and Lower worked quietly. Public attention did not seem of benefit, and, anyway, no one else seemed to be working on heart transplants, not seriously, and so they had the leisure afforded by a lack of rivals. In an era of competition, of the space race (Sputnik had been launched the year prior) and scientific races more generally, neither Shumway nor Lower felt the urge to do great things merely for the sake of being first.

In 1959 Shumway and Lower were ready to perform their first heart transplant out of one animal and into another. In theory, it might be easier than putting a heart back into the same animal from which it had been removed (there would be more tissue to work with). But theory and practice are very different in the laboratory. At that point, Shumway, and with him Lower, had been offered the opportunity to move to Stanford Hospital Center in Palo Alto, California. The position came with strings and caveats, but it also came with a big new lab, and so he took it, and they moved their dogs and ambitions down the road.

In Palo Alto, on the big day, a healthy but wild-eyed dog was chosen to be the recipient. Another was chosen to be the donor. Both were anesthetized. The animals were then cooled. Shumway prepared everything, and Lower got ready to do the surgery. Carefully, Lower cut the heart out of the recipient dog and put it to one side. After that, he removed the heart from the donor, each step taking just a few minutes. He then sewed the donor heart into the recipient dog. Shumway was given the honor of shocking the heart, to bring it back to life. The heart began to beat. The heart-lung machine was unhooked. The whole process took less than an hour; the two men had just transplanted the very first heart. There had been some precedent, when Alexis Carrel and Charles Guthrie had transplanted a puppy heart onto the neck of a dog, but the dog's original heart had remained inside it, and the dog died after just two hours. This was the real thing.

The next day the local newspapers carried the story: "Stanford Surgeon Switches Heart in Dog—It Lives." Shumway was thirty-six years old and Lower just thirty; they were still wild-eyed, ambitious boys. By 1962, they had performed four successful transplants; each of the recipients lived many months.[2] Success followed success for the two. In 1963, Lower left Stanford to run his own laboratory at the Medical College of Virginia. But even across the country, the two planned to spend the next ten years, longer if necessary, working together to perfect their method, dog after dog, so that they might figure out how to make transplantation safe for humans. They still needed to find a way to prevent the recipient body from rejecting the donor heart; a way to keep a heart beating in the new body for decades rather than just hours or days.

To the extent that they compared their work to anyone's, Lower and Shumway likened their project to the quest to land men on the moon (this seems to be a favorite analogy among pioneering surgeons). The race to land a man on the moon was about technology as much as it was about discovery or progress. The heart transplant came to have a similar flavor, one of progress. That the goal was valuable was unquestioned by those struggling to attain it.

But while the quest to land a man on the moon was one model for the race to transplant human hearts, there was also another, one often suggested by journalists who wrote about heart transplants. It came from literature, from a story by Mary Shelley. Shelley lived more than a century before the first attempts at heart transplants, but the ethos of inevitable (and hence good) progress was the same one with which she had grown fascinated. The key moment for her inspiration was in the spring of 1816. Mary and her husband, Percy Bysshe Shelley, had gone with Lord Byron; Byron's then lover, Claire Claremont; and Byron's physician to two adjacent houses on Lake Geneva in Switzerland. Rain kept them confined indoors, where they were left to talk and write.

One night, as the rain pounded outside, Mary was sitting around with these friends telling ghost stories. At first, they read from *Les Fantasmagoria*, a French translation of a German book of horrors, and then Byron proposed that they should each compose a horror story. Mary did not tell a story at first, and it might have stayed that way, the tall tales left to the big boys. Then, on June 21, 1816, she listened to a conversation between her husband, Percy, and Byron and it got her thinking about different sorts of ghosts, those of science and progress. Percy mentioned that Erasmus Darwin (1731–1802), Charles Darwin's grandfather, had written about using science to bring dead animals back to life.[3] Mary listened intensely. The idea was titillating and horrifying. Bringing things to life? It got her mind spinning. In a time before the essence of life was understood, one could realistically imagine Erasmus Darwin surrounded by dead animals that he was trying, one after the other, to reanimate.

In the middle of the night, Mary Shelley woke up, her mind swollen with an idea. She had what she called a waking dream. In it, she saw a pale scientist "kneeling beside the thing he had put together." He was looking down on the "hideous phantasm of a man stretched out," which then, thanks to some powerful engine, "showed signs of life" and stirred "with an uneasy, half vital motion." To Shelley, this scientist mocked the magic of life, the mechanism that lay hidden inside the body.

She would write an entire book about the ghosts of progress; she would write about the ends to which science might go. England was turning dark with technology, industry, and the hope of progress. This, to Shelley, was one of the scariest things she could imagine, a story in which the monster was both the creature set in motion by science and the science itself. Her monster would be built out of the parts of other creatures. It would be assembled, piece by piece, until ultimately it was given the final elements that bestowed upon it life.

In the book, Mary Shelley did not describe just what it was that reanimated the monster, but in the science of the time, only one organ could do the job she required, that of giving biological life and emotion:[4] the heart. The monster Shelley created was Frankenstein's monster. This monster would go on, because of his emotional heart, to look for love and kindness among humans, and then—failing to find either—to terrify the society and the scientist who created it.

For the heart surgeons, comparisons to Dr. Frankenstein's love of progress were unflattering, but nonetheless, they were made almost as soon as Shumway and Lower took the heart from one dog and put it into another. The criticism implicit in such comparisons would continue over the next decade as other surgeons readied themselves for the possibility of performing a heart transplant on humans. One of the most outspoken critics of heart transplantation was Werner Forssmann. Forssmann, the guy who put a catheter into his own heart without even trying the procedure on a cadaver first, was urging moderation.[5] Forssmann warned that without understanding the potential for rejection of organs by the bodies of recipients, it was too soon to contemplate transplants. It was progress without consideration. Yet Shumway and Lower were considering. They were waiting and trying to understand when and why transplants worked; they were trying to understand rejection. In their way, they intended to heed Forssmann's recommendation to wait until everything was just right. But it wasn't up to just them; as time went on, other surgeons began to contemplate transplants, surgeons including Christiaan Barnard.

Barnard was born in a small town in South Africa. He was South African in the time of apartheid, and he thought that both he and his country deserved more respect than either was given. If he was to transplant a heart, it would bring him fame; it would bring his country fame. Barnard wanted to do this the way a little boy might see a fireman and want to be one. He was a good surgeon and a hard worker, but he was not trained in heart surgery, much less in the intricacies of heart transplants. He did not know the history of the heart; he stepped into its story abruptly without taking the time to catch up.

After standard medical training in South Africa, Barnard had gone for graduate training to the lab of Dr. Walt Lillehei at the University of Minnesota, where he imagined he would learn the best techniques in American surgery, techniques he could take back to South Africa. Barnard's doctoral dissertation focused on a congenital intestinal disorder. But he briefly met Shumway while working for their shared mentor. Lillehei had been a pioneer in the exploration of the heart, but he had done so by breaking every conceivable rule. On multiple occasions, he was nearly imprisoned for his actions. In relatively few months of training with Lillehei, Barnard learned three things. He learned how rapidly advances were occurring in heart surgery. He learned that one did not need to follow the rules. And he learned that Shumway and Lower, two men he had never heard of before, were slowly taking the steps necessary to one day transplant a human heart.

Barnard returned to South Africa after his stay in Minnesota and, while there, became convinced that he could and would perform a heart transplant. In 1958, the same year Lower and Shumway first removed the heart from a dog and then put it back, he was appointed as a surgeon at the Groote Schuur Hospital in Cape Town, South Africa. There, despite his relative inexperience with the heart, he established a heart unit. Soon he was promoted to lecturer of surgical research and then head of the division of cardiothoracic surgery. Locally, his star was beginning to rise.

Years passed and eventually Barnard decided he wanted to see the progress Lower and Shumway had made. He visited Lower's laboratory at the Medical College of Virginia (now part of Virginia Commonwealth University) in 1966. Barnard stayed three months with one of Lower's colleagues, David Hume (who had made advances in kidney transplants), during which time he learned rapidly and aggressively. Barnard watched Hume, carefully noting the drugs he used as immunosuppressants. He also watched, with numbed amazement, as Lower performed a heart transplant on a dog. To some men, such an event would have seemed a horror or a miracle, or both. To Barnard, it was a lesson. In watching Lower, Barnard was now sure that heart transplants were ready for humans. He was sure, too, that he could do one. He even confided to one of Lower's assistants that this was his intent. When the assistant told Lower, Lower was unworried. Barnard, after all, knew almost nothing about heart transplants. How could he possibly contemplate doing one?

At this point, Barnard's brother, Marius, joined his program in Cape Town; he would work alongside Christiaan. When Christiaan Barnard returned to South Africa after his trip to Virginia, with Marius at his side, he was ready to do a heart transplant. He wanted his hospital and South Africa to be first. He began to prepare everything for such a surgery. This would take more work than it might have elsewhere, for the simple reason that in South Africa, Barnard did not have the resources other hospitals had. He did not have a team of cardiologists. Nor did he have all of the right equipment. For example, he did not have an autoclave large enough to sterilize the big equipment. He would make do.

What Barnard needed most was a patient in need of a heart, and a body that could donate one. Then a patient came to the hospital who clearly would not live long with the heart he had. Louis Washkansky was in bad shape, and when Barnard explained the possibility of a heart transplant to Louis, he (and eventually his wife) agreed to the procedure. Louis's wife, Ann, asked Barnard pointedly what the chances of her husband's survival were, and he responded, "Eighty percent." This despite the reality that in the transplants that Christiaan Barnard and his brother had by then attempted on dogs, nearly all of the recipient dogs had died during the procedure, and none of those that survived had lived more than a week.

What was next was simply to wait for a donor to turn up. Until this point, the donors for other kinds of transplants had all been individuals who were dead, but Barnard was ready to consider something more radical, donors whose bodies were still living but whose brains were dead. This improved the odds of finding a donor dramatically because one did not have to "capture" a heart in the seconds after death. But the idea of using brain-dead patients (patients that would, with time, come to be called "living heart cadavers") pushed heart transplants onto even newer ethical ground.[6] Barnard was ready for new ground. In much of the world, the law clearly stated that death occurred when the heart stopped, and so the hearts of heart-transplant donors needed to actually stop before they could be used, but in South Africa, the law was more ambiguous. If Barnard could find a donor, he would not have to wait for his or her heart to stop—all he needed was for the donor to be brain-dead. He just had to wait for someone in this condition to appear; he just had to wait for someone's brain to die.

By the fall of 1967, with Washkansky's condition deteriorating, Barnard felt everything was in place to do a transplant. He didn't yet know how to deal with the problems that might emerge in the recipient's body. He had done little in the way of experimentation, and yet he knew that others were getting ready to do their own transplants, and so if he wanted to be first, now was the time. The increasing successes of Lower and Shumway with dogs were being publicized. Shumway had announced his readiness to perform a human heart transplant. Adrian Kantrowitz at Maimonides Medical Center was ready too, just waiting on bodies, one in need of a heart and one with a heart to give. Two surgeons in Texas, Denton Cooley and Michael DeBakey, were beginning, separately, to consider the possibility of doing transplants. Donald Ross, a former classmate of Barnard's at Cape Town University, was at the National Heart Hospital in London, and he was ready too. What was more, in Mississippi, James Hardy, a talented surgeon, had already done something astonishing. Whereas most surgeons focused on transplanting organs from one human into another, Hardy had another idea. He purchased four chimpanzees and had them shipped to the University of Mississippi Medical Center, where they were cared for as Hardy waited for a patient in need of a heart transplant. Then one arrived, Boyd Rush. Rush was comatose and had only a weak pulse. His left leg was gangrenous and his face was pocked with blood clots. His heart was unable to move enough blood through his body; it was failing, and it apparently had been failing for a while. Rush appeared to have just hours or, at most, days, to live. On January 22, 1964, Hardy amputated part of the man's left leg and readied him to receive a chimpanzee heart. Later that same day, Hardy opened up Rush's chest and pulled out his heart. It was, Hardy would later say, "an awesome sight," the empty space where a heart should be inside a still-living body. The next step was to stitch in a chimpanzee's heart, which Hardy proceeded to do; it took him nearly an hour. Hardy's surgery, we know in retrospect, had very low odds of success. Rush was in terrible condition, but more importantly, he would almost certainly have an immune reaction to the chimpanzee heart. Yet the experiment worked. After an initially unsteady beat, the chimpanzee heart beat in the man's chest for ninety minutes. Rush was, for those moments, alive with a chimpanzee heart inside him, though he soon died of unrelated causes. Hardy's experiment horrified the public—people questioned the ethics of transplanting chimpanzee parts into humans. But with time, it would also serve to embolden the other surgeons, the men standing, in the fall of 1967, on the precipice of trying the same with a human donor heart.

If Barnard was going to perform the first human-to-human heart transplant, he was going to have to hurry. Three years had already passed since Hardy's surgery, and nine years had passed since the first heart transplant in a dog, years during which, it seemed, ever more surgeons had had the time to consider the basics of the procedure. On November 22, Barnard received a call about a potential donor, but an EKG suggested that the man's heart might be damaged, though it appears the man's race—he was black, the prospective recipient was white, and this was apartheid South Africa—also played a role in the decision to decline his heart. Then, on December 3, 1967, Barnard got another call. A twenty-five-year-old woman, Denise Darvall, and her mother, Myrtle, had been walking across the road after having bought sticky caramel cake at their favorite bakery when they were hit by a truck driven by Frederick Prins, a police reservist who had had too much to drink. The impact killed Myrtle and critically injured Denise. By pure chance, Ann Washkansky happened to be driving down the same road and saw the scene of the accident. She shuddered at the tragedy, not yet understanding the dark complexity of its consequences. Denise Darvall was taken from the scene directly to Groote Schuur Hospital so she might be resuscitated, but that was not possible. It was decided that her brain's life could not be saved, but her heart's could. Machines were set up to keep her body alive as long as possible.[7] Her heart beat normally, pumping blood to every one of her functioning organs, and even to her damaged brain.

Barnard raced to the hospital. Once there, just to be on the safe side, legally, he, his brother, and a colleague (all of whom swore never to speak of it again) surreptitiously administered potassium to Denise's heart, which temporarily stopped it, allowing Barnard to say that her heart had stopped before he removed it. It was 2:20 a.m.; it was going to be a long night. Denise's sternum was then cut with a saw and Marius Barnard connected her to a heart-lung machine. There she waited, open. Denise's body and heart were cooled to 28 degrees centigrade to maximize how long the heart could wait. Next, in another room, Washkansky's heart was removed, and he too was connected to a heart-lung machine and cooled.

Then Barnard went back to the first room and removed Denise's heart, placed it in a small container, and carried it to the room where Washkansky and his open chest waited. He lifted the small heart up and placed it into the great cavity in Washkansky's chest, where he stitched it to Washkansky's arteries and veins. At 5:43 a.m., Barnard undid the clamps and allowed the blood from Washkansky's body to flow into Denise's heart. The heart turned pink and, after some more interventions, at 6:13 a.m., Barnard announced that it was time to turn off the heart-lung machine. The heart beat normally, and as it did, Barnard's heart raced. It was going to work.

The next day, December 4, the result was announced in the South African newspaper the *Star*: "Transplanted Heart Is Beating!" Who knew South Africa had a heart-surgery program? (It didn't; it just had Barnard and his brother.) Barnard's achievement was on the front page of almost every paper in the world. As Barnard made his way home, calls had come in from France, London, and nearly everywhere else. By Monday, CBS and the BBC had news teams in Cape Town. Everyone knew about it. Washkansky could speak; he could eat breakfast. Although he now had the heart (and, many would say, soul) of a twenty-five-year-old woman beating in his body, he was still the man he had been. Washkansky's wife feared that with a new heart, he would no longer love her, but it seemed he loved her just the same, perhaps more. On December 15, Washkansky was on the cover of *Life* magazine, smiling, slightly. On the same day, Barnard was on the cover of *Time*, a drawing of his head and shoulders in front of an illustration of a heart, as though the heart had been transplanted into him. Barnard would go on to live a new kind of life. He was famous. He began to describe the "years of animal research that it took to build up to performing a human heart transplant," failing to mention that that research was done by Shumway and Lower, not himself. He socialized with and slept with movie stars. He traveled the world. The *New York Times* noted, "This is one of the peaks of modern scientific achievement, fully comparable to the heights scaled earlier in such fields as space exploration or modern biology." Barnard bathed in the praise. He loved it.

As for the actual heart-transplant recipient, Washkansky lived another day and then another. It was a seeming miracle. A week went by, and then two, but then things took a turn for the worse. By day fifteen, Washkansky's immune system, it appeared, had started to react to the transplant and to attack Washkansky's own lungs. Washkansky was given massive doses of drugs that would suppress his immune system, Imuran and prednisone, but at great expense. Bacteria already dwelling in his lungs, both klebsiella and pseudomonas, began to grow unchecked. By day eighteen, Washkansky was dead.

Here is where the story of heart transplants departed tragically from that of the moon landing. The first moon landing was an unqualified success. Not so with the first heart transplant. Yet, after a brief period of sadness, Barnard continued with his celebration. Just a few days after Washkansky's burial, Barnard was sitting in first class on a plane drinking champagne en route to TV interviews in New York and Washington, DC. He traveled the world giving talks and interviews. He went to Hollywood parties. Meanwhile, several groups of surgeons who had spent years preparing to do what Barnard had done[8] continued the race to transplant hearts themselves. From the perspective of long-term survival, there had still not been a successful transplantation of the heart.

Days after Barnard's surgery, on December 7, Adrian Kantrowitz performed a heart transplant on an infant. Kantrowitz had been ready to do a transplant a year and a half before Barnard, but when an opportunity came, the donor heart was in poor condition, and so he had decided to wait. After several near misses, a healthy donor heart arrived when he also had a needy recipient. Kantrowitz had performed heart transplants on more than four hundred puppies. Other than Shumway, he was the best prepared in the world to perform a heart transplant. But the recipient infant lived just six hours.

Kantrowitz, in a moment of dejected humility, pronounced the surgery an "absolute failure." He performed a second operation, with the same result, and so, after hundreds of puppies, great patience, and a decade of getting ready to perform that heart transplant, he gave up on the entire field.[9] A month later, Shumway transplanted a heart into a man named Mike Kasperak; Kasperak lived just fifteen days. In Texas, Denton Cooley would perform seventeen heart transplants in the next year. All of the recipients died within the year, though in many cases, not before they had a chance to talk to the press. The race to transplant hearts had become a deadly circus. Barnard had sped everything up—everyone credited him with that—but at what cost?[10]

By December 1970, a mere three years after the first heart transplant, 175 transplants had been performed. Just twenty-three recipients (including Barnard's second patient) were still alive; perhaps, some speculated, more patients would have been alive if none of the transplants had been performed in the first place, if those in need of hearts had simply been left alone.[11] Most had died in the days or months following the surgery. Here was an amazing, awe-inspiring surgery. It was a technological miracle, but one that created lives that, like Frankenstein's monster's, were fated to end badly. No one had yet figured out how to keep bodies from rejecting donor hearts or how to reliably ward off infection, so except where good luck and fate intervened, hearts could be transplanted into people who needed them, but the donated hearts and their recipients would soon die. In competing, the early heart-transplant surgeons were pushing a field faster than it was capable of going.

By this point, Lower and Shumway had spent longer than anyone else working toward human heart transplants. They had also performed more heart transplants than anyone, and had a much higher rate of success than anyone (about 42 percent of their patients lived more than six months, whereas less than 10 percent of heart-transplant recipients in general did). But even in their hands, more than half of recipients died within six months. And so when Lower performed another transplant and his patient, like so many others, died, it was not surprising. What was surprising was that, in the wake of that death, he found himself accused of murder.

On May 25, 1972, in Richmond, Virginia, Lower had attempted another heart transplant. It succeeded for a short time and then failed, but it was not the recipient's death of which he stood accused. It was the donor's.

Bruce Tucker had fallen onto the concrete at the egg-packing plant where he worked. The fall resulted in severe brain damage. Tucker was brought into the hospital, where doctors tried to relieve the pressure on his brain with a craniotomy, but there was no response. Tucker was placed on a respirator, but he seemed to be only "mechanically alive." The next day, the hospital declared Tucker "unclaimed dead," and the surgeons bent over him, cut his chest, broke his ribs, and opened him up. Once the chest was open, they cut out his heart and lifted it up carefully so that it might be placed in the body of another man. In the eyes of Lower, removing Tucker's heart was the first step in saving another man.

This was not how Tucker's family saw things.

According to Tucker's family, Lower and his team took Bruce's heart before he was dead; they took the heart even before his family had been found. In the family's eyes, Lower was just waiting for a heart. He already had the body of a recipient, Joseph Klett, laid out and ready. Tucker, a fifty-four-year-old African American man, just had the bad luck to be black and show up at the hospital when Lower was waiting for a heart. The hospital did not, it seemed, expend much energy searching for Tucker's family before pulling him off life support and declaring him dead. Tucker's brother, a shoemaker, called the hospital repeatedly searching for his brother and was told, variously, that he was in surgery or that he was in recovery. There was no mention of a heart transplant. When a friend called the hospital, the friend was told that Tucker was not even there. Even when Tucker's brother arrived at the hospital after the heart transplant had taken place, he was still not told of his brother's fate. The state had a mandatory waiting period of twenty-four hours for organ harvests from brain-dead patients, which the hospital seemed to have ignored. To Tucker's family, Lower was a modern Dr. Frankenstein, waiting for parts, ready to gather them at any cost.

Tucker's family hired a hungry young African American lawyer, Doug Wilder, to argue their case; Wilder would later become the governor of Virginia.[12] He had big ideas for his own future, and, perhaps with those ideas in mind, he made even bolder statements than the family themselves felt comfortable making, statements that played on the emotion of the Tucker's story. He suggested that, given another day, Bruce Tucker might have started to recover. Lower, he said, killed the poor man. Lower killed him as surely as if they had met in a back alley. He had taken the most precious thing he had: his heart.

Personally, Wilder could not possibly have been more prepared to work for the Tucker family. Before Lower's surgery, Wilder argued against heart transplants, particularly the hearts of African Americans into whites. He wrote, "They're not going to be taking the hearts of any white mayors. You know whose hearts they are going to be taking." Even before Lower's surgery, the discussion of heart transplants was racially charged in Richmond. Then, under questionable circumstances, Lower took the heart of a black man and gave it to a wealthy white man.

At stake in this case was not only Lower's future but also the future of heart transplants. A neurologist had seen Tucker after his fall, but he had not pronounced him brain-dead. He had, more casually, said that it was "very unlikely" that Tucker's status would change. The neurologist left open some chance of hope, however remote; it was a verbal loophole that allowed Tucker's family to imagine that he might have recovered if only he had been left alone.

*   *   *

Much depended on how one determined if a person was dead. For most of the history of medicine, beginning in ancient Greece, life ended when the heart stopped. To Christians, the Bible made clear that "the life of the flesh is in the blood." God breathed life into man's nostrils, which suggested the circulatory system's activity was synonymous with life. Similar sentiments exist in the Torah and the Koran. In the United States as well as in many other countries, heart death was legal death, prior to the advent of transplant surgeries. In recognition of this history, the Americans who had been readying to do heart transplants when Barnard beat them to it assumed, for the most part, that they would have to wait for a donor heart to actually stop, for the donor's brain and heart to die, before the heart could be harvested. Barnard, in moving forward, changed the debate, even if ambiguity about just whether and how Denise Darvall's heart had stopped persisted. However other surgeons felt about Barnard's achievement (and many did not feel terribly good), most credited him with changing the discussion. The heart, so long the seat of love, passion, the soul, and even thought itself, was becoming just another organ.

In response to Barnard, a dean at Harvard University convened experts to write what came to be called the Harvard Code, consisting of criteria that defined *life* as "life in the brain" and *death* as "brain death." The Harvard Code recognized that a body without a heart could be returned to life with a heart transplant or perhaps even an artificial heart of some sort, but a body without a brain could never and would never be. *Brain death* was defined as the point at which a patient became unaware of his surroundings; was unable to move spontaneously; and had no activity on a brain tracing, an electroencephalogram (EEG), which measured the brain's electrical activity. In response to the Harvard Code and changes in how surgeries were being done, many hospitals began using brain death as the criterion for the end of life. But any such code is ultimately cultural, a definition of life with fuzzy boundaries, especially until it is made law, which, in Virginia at the time of Lower's most recent surgery, it had not been.

Lower was initially optimistic about the trial. He also felt good about Tucker's role. To his mind, he had done nothing another surgeon would not do in the same situation. Then the judge made an early and seemingly fateful decision. He asked the jury to define life as ending when the heart stopped beating, not with the death of the brain,[13] based on the definition of *death* in *Black's Law Dictionary*. To Lower and to heart surgeons around the world, this was a blow. If the beating heart was, as in ancient times, the defining feature of life, then surely Lower had killed a man, but then the same guilt would fall upon other surgeons who had performed heart transplants (if the heart was allowed to briefly stop, everyone was happier, though this was really a sort of technical work-around more than a reasonable distinction, a work-around of which Barnard had taken advantage). Lower knew the heart was just an organ. How could the judge make such an argument? The academic part of Lower's mind wanted to argue, to plea on behalf of reason. The rest of his mind had begun to scream. It had begun to contemplate the possibility that because of his attempt to move medicine forward, he might end up in prison for life. An article in the *National Observer* noted that because of the use of the definition of *death* in *Black's Law Dictionary*, "The jury would almost certainly be compelled to convict the doctors."

Tucker's family, who initially thought they faced an uphill battle, began to feel as though they might win. To the family, broader narratives compounded the treatment of Bruce Tucker. African Americans have a long history of terrible, tragic, amoral, and immoral mistreatment by the medical system. The family sought $900,000 for deprivation of civil rights and an additional $100,000 in wrongful death. Whatever one considered the end of life, they seemed to have a case. In this light, when Wilder told the jury that Tucker was just another one "of the faceless blacks to the [hospital] hierarchy," many in the black community heard a truth being spoken.

Recent legal precedent was on the family's side too. Two and a half weeks after Lower's team removed Tucker's heart, Japanese surgeon Juro Wada had carried out Japan's first successful heart transplant. The recipient, an eighteen-year-old boy named Miyazaki Nobuo died from a lung infection eighty-three days later. Upon the news of the patient's death, another Japanese doctor at Sapporo Medical College accused Wada of having killed the donor, a man who was declared brain-dead after a swimming accident. Eventually the case against Wada was dropped, but not until long after Lower's case had concluded, and not because he was seen as innocent but because the evidence of the medical status of the donor was too limited.[14] At the time of Lower's trial, it was not yet clear that Wada would win, though it was clear Wada's career in heart transplants was over. Lower's might be too.

Lower became increasingly anxious. His lawyers spoke on his behalf, but quietly, modestly. Then the judge made another announcement: the jury, he decided, could consider either definition of *death*. That is, they could choose to consider the possibility that death was, as Lower would have it and in line with the Harvard Code, defined by the "complete and irreversible loss of all function in the brain." It is unknown just what made the judge change his mind[15]—he seems to have been influenced by the consensus that had emerged among American surgeons—but once he did, he set the stage for Lower, Hume, and their colleagues to have more of a chance to win.

As Lower waited to hear the verdict, he was aware that he was not the only one on trial; his failings, if that was what they were, were failings of a field, failings of technology and unchecked progress. Heart transplants were on trial, as was the heart itself. For thousands of years, the beating heart had meant life. Doctors had stopped talking about the soul, but they still talked about life, and for doctors, life had moved to the brain. But in this courtroom, a jury of men, their own hearts pounding and thumping, would decide whether this was still true. If it was not, if life still resided in the heart, then every heart-transplant surgeon who had harvested a heart from a brain-dead patient was a murderer.

The jury passed slips of paper over to the bailiff, who passed them to the judge. The judge's big hands unfolded the papers. He did this as slowly as he could, it seemed to Lower. Then he started to announce the verdict. Lower looked up. The dry lips of the judge opened, and he said, "Not liable." In the courtroom of an aristocratic judge who styled himself as part of Old Richmond, an all-white jury had sided with the white doctor. Lower sat up straight and began to cry. The Tucker family members sank down into their chairs and began to sob. Their son and brother was dead, their case lost. In legal contexts, Lower's name became associated with the timing of death. Death, after the *Lower v. Tucker* case, came to be viewed as brain death.[16] The case would be revisited[17] as state laws were changed to consider brain death as death, but the general societal lesson has never changed. In Richmond, consciousness and the soul were banished from the heart.[18] The legal system had taken a bold step. But contrary to their natures, the surgeons pulled back. Heart transplants became taboo—not immediately, in response to Lower's case, but in the years to follow, years of reckoning.

For surgeons, the residual possibility of legal problems remained, particularly in those states and countries in which laws clarifying the meaning of death had not been passed. The bigger problem was that, most of the time, the surgery did not work. People were reborn with new hearts, but they did not survive. They sat up, smiled, hugged their loved ones, and died. In many cases heart-transplant recipients still seemed to die sooner than they would have without the new hearts. This was enough to dissuade surgeons from continuing. More cynically, one might argue that after the first heart transplant in the world and then the first in Japan and the United States, the excitement of subsequent surgeries was diminished; what was left was the far less glamorous task of figuring out how to keep the heart from being rejected by the recipient's body. Whether that task could be accomplished was unclear, and even those surgeons who performed multiple heart transplants eventually stopped. Dr. Michael DeBakey of the Methodist Hospital in Houston performed twelve transplants but then stopped. Dr. Denton Cooley of the Texas Heart Institute in Houston at one point led the world in the number of heart transplants but then stopped. Christiaan Barnard performed ten heart transplants, then stopped. In 1971, a cover story in *Life* magazine predicted heart transplants would be abandoned, and, when they were, it would conclude "an era of medical failure."[19] Whereas there were 121 heart transplants done in 1968, there were just 47 in 1969, 17 in 1970, and about 10 in 1971.

Here and there, other surgeons tried transplants, but most abandoned the procedure as quickly as they had started. Shumway and, with him, Lower persisted. The two (but particularly Shumway) came to dominate the field in the way it had long been imagined they would.[20]

Shumway was the first to admit that, even for him, the surgery was more often fatal than successful. Shumway had watched as, one by one, his heart recipients died (albeit after more months and years than for other surgeons). Some died because the hearts they had been given were too big or too small. Others died because their own bodies or their donor hearts were too far gone. Most died because of infection or because their own bodies ungratefully rejected the hearts they had been given.

Shumway worked to resolve the problems of infection and rejection; they were two sides of the same coin. Infections were caused by pathogens introduced during surgery or later, and they spread readily because the medications, such as steroids, used to suppress the immune system suppressed it too much and too generally. Shumway would use a series of compounds and approaches that would calm a recipient body enough for it to accept a donor heart but not so much that it would allow invasion by pathogens.[21] It was not exciting science. Shumway would not be in the news again for his efforts. Yet it was the science necessary to change heart transplants from a sideshow novelty into medicine. For this, he received little credit. Even his obituary in the *New York Times*, before mentioning that he was the man who really made heart transplants possible, noted that it was Barnard who performed one first.

The solution to the problem of rejection would be twofold. Shumway realized early on that transplants of hearts for which the blood types matched had a better chance of avoiding rejection. This would never have been resolved through the study of dogs alone, because dogs do not have as many blood types as humans. Still, even when blood types were matched, bodies rejected the new hearts they were given. Shumway needed to suppress the reaction of the immune system to the new organ; Barnard knew this too, of course, but he chose to disregard the problem's complexities; rather than trying to circumvent the immune system, he just hoped to get lucky. Shumway hoped for nothing; Shumway planned and tested.

The immune system's primary job is to differentiate *self* from *other*, *us* from *them*, and then to respond accordingly with war or peace. Not all others are attacked. Your body, for example, takes care of some of the bacteria on your skin and in your gut; it takes care of the many species on which you depend. But organs from other bodies are rejected, at least initially, and the more different they are, the more likely they are to be rejected. Shumway believed that if this response was pushed off for a while, the immune system might come to think of a new organ as part of the self; us, not them. The key was suppressing the initial immune response.

Shumway attempted many procedures to suppress the immune system, each resulting in rooms full of dead dogs with broken hearts. Then, in 1971, the same year that *Life* magazine had predicted the end of heart transplants, came the first steps of a breakthrough. While on vacation in Hardangervidda, a high and desolate plateau in Norway, a Swiss researcher at Sandoz (now Novartis),

Jean-Francois Borel, decided to search in the soil for organisms capable of producing useful compounds.

Borel was particularly interested in finding antibiotics, and soil bacteria and fungi are a rich source of antibiotics. The first samples Borel tested from a fungus he collected did not seem to produce useful antibiotics. But following company protocols, he also tested the fungus for other effects. In these later tests, some extracts from the fungus seemed to alter the behavior of immune cells grown in petri dishes. The same extracts also suppressed immune function in live mice. Borel figured out that the active compound in those extracts was cyclosporine, but the cyclosporine did not suppress all of the immune system. It suppressed just that part of the immune system whose job it is to distinguish self from nonself: the T-helper cells. It suppressed precisely that part of the immune system that causes problems for transplanted hearts; this, Borel quickly realized, was a very big deal, or it might be, anyway.

In 1973, two years after Borel discovered cyclosporine (four years after it was first collected in the field), Shumway was described in a newspaper article as a man who "still dreams of the day when the biochemists will produce an exquisitely precise antirejection drug that will protect transplanted hearts, kidneys and other vital tissues and yet leave unweakened the body's ability to fight infection."[22] Shumway knew nothing about cyclosporine; news of it had not yet left Borel's lab. The road from the discovery of a drug to its use, production, and approval is long and slow; in the case of cyclosporine, it took twelve years.[23] Borel's first publication on cyclosporine appeared in 1978. In 1980, cyclosporine was used experimentally in humans, and finally, in 1983, it was approved by the FDA after two clinical trials.[24] In the intervening years, Shumway had perfected nearly every other element of heart transplantation he could. Cyclosporine was the last piece of the puzzle, and, almost immediately upon its approval, it became a key compound used during transplants. The use of cyclosporine is not without its challenges.

Even when taking it, most patients require additional immunosuppression (often in the form of steroids). Initially, there were struggles with dosages. Also, the long-term use of cyclosporine and other immunosuppression drugs is associated with a variety of potentially serious health problems. Yet cyclosporine makes transplants that would otherwise be impossible, possible. Thanks to cyclosporine, the number of heart transplants started to accelerate again, dramatically (to a point). By 1987, there were four thousand people in the United States with transplanted hearts, many of whom would not have been alive without the transplants, and many of them Shumway's patients. Shumway's quest to fight rejection and Borel's vacation in Norway had come together to yield a major breakthrough that saved lives. Here was Shumway's victory lap, the proud conclusion to his story: the rebirth of the transplant as a real solution for desperate, brokenhearted folks. Tens of thousands of people walk around today with hearts donated by other people. They are true chimeras, and their existence is made possible by the surgeons and researchers who struggled for progress, but also by the patients who lost their lives in the process, patients such as Louis Washkansky, who gave everything for what might be.

As is the case with most medical discoveries of novel compounds, no one ever went back to consider just what cyclosporine does in the natural world. Why does a fungus produce such a potent immunosuppressant? Recently, a clue has emerged, thanks to a group of undergraduate students and two fungus biologists (mycologists).

The discovery began when the mycologist Kathie Hodge, a professor at Cornell University, decided to look at two specimens of fungus collected by students who'd taken a fungus course taught by Cornell professor Richard Korf in the fall of 1994 at Michigan Hollow State Forest in Danby, New York. The specimens were small, each composed of white-tipped stalks of fungus crowned by yellow fungal "fruits" (perithecia). The stalks flowered out of the white tumescent bodies of scarab beetle larvae. The beetle larvae looked as though they had been living in piles of shit, a common place for many beetles.[25] Kathie Hodge decided to try to identify these unusual fungi.

Hodge is a preeminent expert on fungus, and to her, the fungus looked weird. It was the sexual phase of the fungus, she knew that much. But when it came to identifying it, she had trouble. It looked to be a representative of a group of fungi, *Cordyceps*, known for altering the behavior of insects. When everything goes right for them, *Cordyceps* fungi land on an insect's body. They then grow through the exoskeleton into the body cavity and up into the head, where they alter the insect's behavior and cause it, in some cases, to climb up into the branches of trees. So elevated, the fungus grows out of the insect head, produces a reproductive structure, and waits to be dispersed through the air. Such fungi are common and diverse. They differ in their particulars and yet share the ability to flower out of the bodies of living insects. In some forests, if one looks carefully, whole colonies of ants can be seen with their mandibles dug into leaves and stems, fungi growing tall out of their heads. To Hodge, the fungi in the beetles looked to be a *Cordyceps*, but which one?

Hodge cultured the dry samples, and when she did, they looked like another sort of fungus with which she was familiar—a kind that had never before been linked to *Cordyceps*. It looked like the fungus from which cyclosporine was isolated. It was. Cyclosporine had, it turns out, been isolated from the asexual stage of the same fungus Hodge found in the beetles, and as is often the case with fungi, the asexual stage and the sexual stage look so different (much as we look different from our sperm and eggs) that each had been given its own species name. Hodge realized the two species were just the sexual and asexual stages of the fungus *Cordyceps subsessilis* (it was later renamed *Eucordyceps subsessilis*).[26] Suddenly, the Norwegian fungus had a more complicated story. It lived in New York too, and it lived in beetles. It seems possible that cyclosporine is produced by the fungus in order to get past the immune systems of beetles and take over their bodies, much in the way that a transplanted heart needs to get beyond the immune system of the recipient. The real innovation for suppressing the immune system had come from evolution's innovation, just as with the use of penicillin to fight bacterial infection.

It was thanks in part to the ancient story of the beetles and fungus that heart transplants eventually became an honest form of medicine. Annual sales of cyclosporine are now in the billions of dollars. More important, heart transplants now number in the thousands. In 2012, more than thirty-five hundred heart transplants were performed. Eighty percent of heart-transplant recipients now live more than a year, 77 percent more than three years, and 70 percent more than five years. One man, Tony Huesman, lived for thirty-one years with a donated heart—eleven thousand days of extra life: eleven thousand breakfasts, eleven thousand nights of sleep, eleven thousand mornings. Heart transplants are as routine as a procedure that requires one person to die so that another may live can be. Though just how many are performed seems to depend as much on culture as on medicine or science. Two-thirds of all heart transplants in the world are done in the United States, where much of heart-transplant research was pioneered and where both court and cultural cases favor brain death as the end of life. This is true even though fewer than one in twenty of the roughly one million people who need heart transplants live in the United States.

For those who get heart transplants, the surgery is a miracle. For this, there are many to thank, including Hardy and Barnard, and especially Shumway and Lower. But if one looks at the big picture of lives lost and saved and money spent (the average heart transplant costs a million dollars), success seems more complicated to define. At any moment, tens of thousands of people's lives could be extended through heart transplants. Thousands more would be saved if more people donated their organs, but there will always be too few hearts, tens of thousands or even hundreds

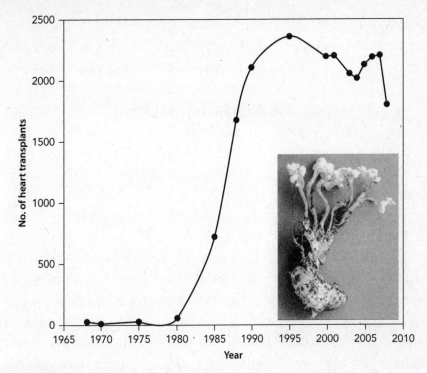

The trajectory of heart transplants. The first surgeons to perform heart transplants hoped they would one day be routine, but this routineness was delayed by the wait for immunosuppressive drugs and then later, until the present day, by the limited supply of living human hearts. Inset image shows the beetle larvae in which the fungus that produces cyclosporine grows. *(Courtesy of Kathie Hodge)*

of thousands too few. Early on in the race for heart transplants, it became obvious an additional solution was necessary. Maybe it would be better to build directly on the mechanical innovations of Gibbon and, later, Greatbatch; maybe one could build a new heart from scratch.

8

Atomic Cows

Man is no more than electrified clay.
— PERCY BYSSHE SHELLEY

Michael DeBakey was a pioneer in open-heart surgery, a demanding genius of ingenuity, a perfectionist, a man who believed there would be time for sleeping in death. He wanted each of his waking hours to be spent intensely, achieving. He demanded no less of those around him. By the time DeBakey died, in 2008, he had conducted, by his own count, more than a hundred and sixty thousand heart surgeries, many of them the first of their kind. He was neither polished nor handsome. What marked him was his ferocious intent to revolutionize whatever lay before him, and also whatever was to the side.

Much of what DeBakey did pointed him squarely in one direction: the building of an artificial heart, a machine of a thing that might tick for years or even forever inside the body of every person suffering from any of the heart's myriad diseases. When he thought about artificial hearts, he did not think of them as scientific novelties. To him, as to many other surgeons, they were the future, the secret to the longevity of millions, maybe billions, of women and men. His was a hopeful, bionic vision of our future, one in which the problems of the heart might be solved through technology and time.

DeBakey had been part of the race to transplant hearts. He and his onetime collaborator and eventual archrival Denton Cooley[1] vied with each other for the fourth and fifth heart transplants, respectively; these two men would go on to vie for much in life. Cooley was the handsome, smooth, well-coiffed, and cheery Texan to DeBakey's, well, opposite. But DeBakey did not want to replace hearts with other hearts; he wanted to replace hearts with artificial hearts, little machines elaborately constructed by surgeon-tinkerers the way that a clock maker might build clocks, clocks out of which entire human lives, rather than cuckoos, emerge. DeBakey became, in addition to one of the world's leading surgeons, the man in the machine shop pounding out the metal, welding the bits, fabricating a version of what nature had spent millions of years carving out of cells. More accurately, he was the man who hired whole teams of men to work in the machine shop. He also solicited the donations to build the machine shop.

The idea of replacing a broken part of the human body with an artificial one is ancient. In one fifteenth-century Egyptian tomb, close inspection of a man's foot revealed an artificial big toe made of leather and wood. *Rig Veda*, an Egyptian text written somewhere between 3500 and 1800 BC, speaks of the warrior queen Vishpala who lost her leg in war and had it replaced with an iron prosthesis good enough to be worn into battle. The Greeks built iron hands, wooden legs, and much more.[2] In ancient Rome, Galen is said to have produced an artificial eye. But the heart is, of course, different. It is one thing to replace a toe, quite another to do the same for a beating muscle, a muscle responsive to the body's temperature, activity, and emotional state.

Some surgeons focused on artificial hearts that were, in essence, extensions of the heart-lung machine, large external devices intended as stopgaps. But DeBakey imagined something different. He wanted to build a small artificial heart that could be placed inside the body and would last for decades, even centuries. Others had dreamed similar dreams in the past. In 1937, Vladimir Petrovich Demikhov invented a device that, when squeezed, could replicate the ventricles of a dog's heart. The device was, by the accounts of those who saw it, amazing, and yet neither a practical solution nor a well-documented one.

When DeBakey began to dream of an artificial heart, between seventeen thousand and fifty thousand patients a year could reasonably be considered recipients;[3] that was a big market, an army of humans who might live years or even decades rather than months. But DeBakey needed more money. The whole field needed more money. DeBakey visited with his famous patients and asked for cash, but he also went to Congress, where he and other scientists pleaded for the development and funding of a federal program to produce an artificial heart. As Joshua Lederberg, a Nobel Prize–winning professor of genetics at Stanford University, put it, an artificial heart was a system "about as complicated as a guided missile or a subsonic bomber," both of which had already been built. The next year Congress responded by forming the Artificial Heart Program within the National Heart Institute (itself within the National Institutes of Health), the first targeted research program in the institute.[4] The program was viewed as so important that it was given a direct line to Capitol Hill. In 1965, the National Heart Institute asked six contractors to come up with plans for an artificial heart.

Even very early on, DeBakey understood the problem with mechanical hearts would be energy, just as it had been for pacemakers. A real heart is fed by food. Food fuels mitochondria. Mitochondria produce energy, constantly, to stoke the heart's contracting cells. Technology had no equivalent, at least none that might last the years one would hope for in an artificial heart. The only options anyone could think of were hearts that plugged into electrical sockets (which was not the sort of grand future people envisioned) and ones that ran on batteries. Batteries seemed the better of the two options, but, given the technology of the time, they wouldn't last long. Even Greatbatch's lithium batteries would be short-lived when asked to run an entire mechanical heart (rather than just signaling a real heart to beat). Surgery after surgery would need to be performed to replace the batteries and restore power to the heart. Then came an idea of the times that seemed to solve everything: nuclear power. A nuclear-powered heart might run forever, or at least forever relative to the longevity of the body's other parts.

Nuclear hearts were first proposed by one of the companies the National Heart Institute contracted to come up with ideas, the Thermo Electron Corporation (now Thermo Fisher Scientific). DeBakey almost immediately felt nuclear hearts were the answer, as did the National Heart Institute and the Atomic Energy Commission. But no one thought the people at Thermo were the right ones to make such hearts. They lacked an understanding, it was said, of the challenges ahead.

Glenn Seaborg was in charge of the Atomic Energy Commission at the time. He was not a physician. He was a physicist, arguably among the greatest of his generation. For much of his career, he sought to find new elements in the universe, and he succeeded. In a series of breakthroughs made possible by technology, perseverance, brilliance, and luck, he and his colleagues had extended the periodic table, adding eight new elements. With each discovery, man's understanding of the materials of which the universe is composed expanded. The elements Seaborg's team discovered included one that justifiably bore his name, Seaborgium. It also included element 94, plutonium. In 1940, Seaborg realized that one particular isotope of plutonium (plutonium 239) produced an astonishing amount of energy when hit by a neutron, enough to make an atomic bomb. Soon, Seaborg was recruited to the Manhattan Project, where he helped figure out how to produce more plutonium, a lot more.[5]

The bomb dropped on Nagasaki, Japan, several years later was a plutonium bomb, a bomb inspired by Seaborg's science. With the energy from plutonium atoms, it killed roughly seventy thousand people and wounded more than a hundred thousand. With this bomb, the war ended in Japan. Seaborg returned to the University of California at Berkeley to live a scholarly life, but he was soon called by President John F. Kennedy to direct the Atomic Energy Commission. He was being asked once more to work on plutonium. The Atomic Energy Commission sought to find as many peacetime uses as possible for nuclear energy, and that included the development of a nuclear heart.

Working with the NHI, Seaborg helped figure out how an atomic heart might best be built and by whom. The best approach, Seaborg thought, was to contract with many groups and see which one proved itself outstanding. Six companies were funded to design their own versions of atomic hearts. Using the best design, the United States would build an atomic heart that would allow people to walk around for years, their every action fueled by the decay of plutonium. Never one to shrink from grand possibilities, DeBakey, who worked constantly behind the scenes during the whole process, persuaded the Soviet government to become involved. That he did so at the height of the Cold War was viewed (at least by DeBakey) as a "supreme act of peace."

Congress suggested Seaborg's Atomic Energy Commission and the Artificial Heart Program together should choose which group would be contracted to build the final working model. The members of the Atomic Energy Commission, with Seaborg at the helm, were supportive, imagining a world in which plutonium had many central uses. The Artificial Heart Program was supportive because this seemed like a way forward to DeBakey's dream of an artificial heart. The two groups agreed to the goal of mass-producing artificial hearts in just five years, by 1970. Together, the Atomic Energy Commission and the Artificial Heart Program were funded to the tune of more than $50 million in 2013 dollars (a relatively great sum, though just one-sixth of what was estimated to be needed).[6] Atomic artificial hearts, it seemed clear, would be a reality in the near future.

The first challenge was that, while hearts are complex but predictable, humans, scientists in particular, can be complex and unpredictable. In theory, Seaborg's group (which knew atoms) and the Artificial Heart Program (which knew hearts) were two parts of a perfectly complementary team. In practice, the two groups worked together only fleetingly before turning the quest to cooperatively make an artificial heart into a kind of passive-aggressive war.[7] The animosity began with disagreement about the order in which the different pieces of the atomic heart should be built.

Choosing the steps necessary to build an artificial heart and their order posed a thicket of challenges. One was how to install the heart in the first place. It had to match up with arteries and veins. It also had to resist decay of any kind, as well as rejection by the body's immune system. This was something that DeBakey, working with his team in Texas, had already begun to focus on. The heart also had to move blood and do so with great force. Then there was the issue of power, the issue the plutonium would solve. The Atomic Energy Commission thought all of these problems should be worked on simultaneously with collaboration between the AEC and the NHI. The NHI disagreed, strongly. It wanted to work first on the heart pump (ignoring how it might be powered) and second on the atomic power source. This and other conflicts could not be resolved, and so the AEC and NHI worked separately toward atomic hearts. The NHI contracted five companies to work on the nonatomic elements of an atomic heart. The AEC sought to develop everything at once and eventually hired Westinghouse Electric to do so.

When plutonium decays, the decay releases energy. With plutonium 239, this is an awful lot of energy, but with the lighter isotope plutonium 238, what is released is more manageable. A pill-size amount of plutonium 238—fifty-three grams—might provide energy sufficient to run a heart for decades, but a plutonium heart would require the invention of a wholly new kind of pump, one very different from the native squeeze box of the heart. In such a pump, nuclear energy had to be converted into mechanical energy, which then would be used to power a pump. Seaborg was ready for such a possibility. He had calculated that hundreds of kilograms of plutonium 238 could be produced each year by the Atomic Energy Commission (by thermal irradiation of neptunium) if, or when, it became necessary.[8] Once the atomic heart was built, Seaborg was sure it could be supplied with energy; he was sure it would soon save many lives.

As years elapsed, the large teams attempted to achieve the joint goals of energy conversion and pumping. By 1972, both teams had made progress. The relative success of the AEC effort depended to a great extent on one man, Willem Kolff, who was, by then, collaborating with Westinghouse. Kolff was as serious a researcher as one might hope to involve in the quest for an artificial heart. He was, at the time, at the Cleveland Clinic, where he worked in the same building as Sones, René Favaloro, and Donald Effler. Kolff had already, by this point in his career, invented an artificial kidney from which modern dialysis machines derive their inspiration. (He built the machine out of parts of a downed Luftwaffe plane, the radiator of an old Ford, orange-juice tins, and artificial sausage skins.)[9] But more to the point, he and Tetsuzo Atsuko, a Japanese engineer, had already reported, at a meeting of the American Society for Artificial Internal Organs, on the development of an electric artificial heart made out of plastic with which they'd kept a dog alive for ninety minutes. In the early 1970s, Kolff had reenvisioned this original model with a plutonium-based energy source and created a working version, but in every model, the heart was either too big to fit into a human or too weak to pump blood.

Separately, the NHI team eventually created a twenty-four-ounce device that took atomic energy (from the decay of plutonium 238)[10] and converted it into a tiny steam engine. All of this took three years longer than anticipated yet at least occurred. But there was more trouble. Just as for the AEC's models, each version of the NHI pump and energy source that could be made small enough to fit inside the body was too weak to do the full job of both ventricles of the heart. In this context (and while the AEC continued to try to produce a fully capable replacement heart), the NHI team tried something different. They changed their goal: instead of building an artificial heart, they would start by making a device that would provide assistance to the left ventricle (which sends blood through the arteries out to the body). It was more modest — a plane to London rather than a spaceship to the moon — and yet still useful.

The next step (en route to London) was attaching a small nuclear energy source and a pump to the heart of a cow. They would intentionally damage a cow's heart and then attach the device, at the descending thoracic aorta, leaving the left ventricle; there, the device could take a weak flow of blood (a result of heart failure) and give it more force. The pump would beat in response to the heart's own electrical impulse, thereby eliminating any need to incorporate a pacemaker into the heart; the device would rely on the real heart's natural pace.[11] In February of 1972, their assist device was implanted into a calf with an intentionally damaged heart. It worked. The heart pumped blood as if it were still sound, or at least it did for eight hours, until the inflow tube in the device kinked.

Meanwhile, DeBakey, who had been given his own pool of money by the NHI ($4.5 million), had also made progress in Texas, albeit on a nonnuclear artificial heart. With a nonnuclear heart, the accepted reality was that the heart would be plugged into the wall. There was no alternative, and yet for patients with no other solution, this seemed, at least at the time, like progress.

The details of the use of DeBakey's heart are subject to (angry) debate. Domingo Liotta, an Argentinean surgical researcher in DeBakey's employ, wrote a conference abstract in which he claimed to have inserted the devices that he and a team in DeBakey's group had developed into each of ten calves successfully. The day *after* that claim, he actually performed what is regarded as the *first* successful implantation of a ventricular-assist device in a cow. The pumps were then put in seven additional cows, with other surgeons present, and all but one cow died within hours. It appears Liotta might have been anticipating his own success in his abstract, success that did not quite occur.[12] Then things really got strange.

The Texas Heart Institute's Denton Cooley, DeBakey's one-time mentee and soon-to-be rival, asked Liotta to build another of the devices he had produced for DeBakey (without mentioning anything to DeBakey) and give it to him so that he could implant it in a human. Liotta asked an engineer to build the device, which the engineer, thinking he was doing work for DeBakey, promptly did. Because the engineer did not pass the machine off directly (he left it to be picked up), he made sure to leave a note on it that said it should not be used on a human because it did not yet really work.

Once Cooley had the device in hand, he waited until DeBakey was away at a conference and then, on April 4, 1969, searched for a potential recipient. He found Haskell Karp. Karp, a patient who needed a heart transplant, had not been able to obtain a heart. One donor arrived, but her heart was in poor shape. A second donor never arrived, and so Cooley installed Liotta's device in Karp as a stopgap until a donor heart could be found. Meanwhile, Karp's wife spearheaded a national search for a donor (donors had become ever more scarce as the success of transplants became ever less obvious). She called out, "Someone, somewhere, please hear my plea. A plea for a heart for my husband. I see him lying there, breathing and knowing that within his chest is a man-made implement where there should be a God-given heart."[13]

One day later, after a potential donor heart arrived in Texas in a condition too poor to be useful, Karp died. After Karp was brain-dead, for reasons that remain unclear, Cooley then transplanted a heart into Karp. The many newspapers that carried the story of Karp's artificial heart said that "the surgery was successful" but that Karp died of other causes. It did not mention the later heart transplant, and, in addition, scholars debate whether the surgery was, even momentarily, a success. When DeBakey saw the news, he was furious and, after a bout of screaming rage, did not associate with or talk to Cooley for decades. At the end of his life, feeling generous, DeBakey extended an olive branch, and the two were, for a short period, reconciled. Liotta's heart was never used again.[14]

The DeBakey-Liotta-Cooley heart was not atomic, and yet, even ignoring the issue of long-term power, it still could not replicate a real heart well enough to sustain Haskell Karp (among other problems, the blood in the artificial heart clogged). This should have been a clue that producing an atomic artificial heart was going to prove difficult. Initially, the advances on the atomic heart and ventricular-assist device (whatever its power) had seemed quick enough that goodwill and hope ruled. To many medical researchers, such a heart had once seemed inevitable. In 1964, DeBakey had imagined an artificial heart to be likely in the next ten years. In 1966, Glenn Seaborg openly spoke of a future in which failing hearts would simply be replaced with nuclear ones. The technology was on track for some kind of success, even if it was more modest than had originally been envisioned. But then progress slowed. Cooley's hubris tarnished the perception of artificial hearts and, perhaps, tamped down some of the eagerness to move forward.[15] By 1976, forty-one individual atomic ventricular-assist devices had been tried in cows, but the devices were little more advanced than what Liotta had already tried, and they were not nuclear.[16] In 1979, a *New York Times* article summarized the state of affairs as follows: "Today, after 15 years and an expenditure of over $125 million a clinical, practical artificial heart for humans is nowhere on the horizon."

Some challenges were technical. Others had to do with the failures of collaborating labs and entities to actually collaborate rather than compete. But as the project matured, it took long enough that the perspective of scientists on the idea of an atomic heart, or an artificial heart more generally, had changed. In 1960, there were few rules for how technology could be incorporated into modern life;

there was no FDA, and scarcely anything in terms of informed patient consent. The Atomic Energy Commission was charged with policing itself, for example. The ethics of artificial organs were discussed, but in academic circles rather than legal ones. But by the late 1960s, many had started to become wary of technology and, specifically, of the consequences of implanting technology into the human body. Intrauterine devices had been implanted in millions of women with tragic consequences that were beginning to be realized. Suddenly a groundswell of sentiment rose up to suggest that the devices implanted into bodies needed to be controlled. A new set of rules was developed for implantable devices. The rules called for a ranking of devices from low to high risk. The mere act of ranking made it clear that whether or not the atomic heart was risky in an absolute sense, it was risky relative to everything else that had been proposed. So were artificial hearts in general. With a single bit of regulation, the project went from being regarded as bold and ambitious to being regarded as both literally and figuratively toxic. Slowly, and then quickly, funds for the atomic heart project disappeared.[17] Seaborg returned, for a while, to his role as a professor at the University of California, Berkeley, and appears to have rarely mentioned the project again. The leaders of the Artificial Heart Program moved into other fields. And then there was DeBakey. DeBakey, who helped to initiate the endeavor to build an atomic heart, had decided to focus on something else: he was trying to perfect battery- or electrically powered ventricular-assist devices. This was, although he would never admit it, a far more mundane endeavor—neither a complete heart nor an atomic one. Today, millions of patients each year receive ventricular-assist devices that aid in the beating of their hearts; they connect via a power cord (going through the chest wall) to several large batteries that must be recharged every two hours. In other words, we have not yet surmounted the problem of power; our best batteries are still humble relative to the efficient power that arises from the work of millions of living cells.

Of course, the initial route that DeBakey had charted, with the goal of building an artificial heart, was seductive enough that some continued to try. They were driven by the extent of the problem — 10 to 20 percent of the more than five million patients with congestive heart failure died each year before they were able to get transplanted hearts or find other solutions. No one would ever mention nuclear hearts again, though (and little has been written about the entire episode), and so the entirety of the body of work on artificial hearts since the late 1970s has focused on large devices that plug into the wall.

The first well-planned attempt to implant a total artificial heart came in 1987, sixteen years after it had been predicted to happen. Dr. William Kolff had by then moved from the Cleveland Clinic to the University of Utah, where, as director for the Institute for Biomedical Engineering, he led a team of more than two hundred doctors and scientists in studies of artificial organs in general. There, he worked with William DeVries and, beginning in 1971, Robert Jarvik. Jarvik was not a practicing doctor. In fact, Jarvik could not get into any medical schools in the United States and so traveled to the University of Bologna to be trained, but he dropped out two years into the program. Disillusioned, he returned home to the United States and decided to try for a master's degree in biomechanics. This time he finished, and it was on the basis of that degree that he landed a job working with Kolff as a tinkerer. With Jarvik watching, Kolff and DeVries installed the Jarvik-7 heart (Kolff had a tendency to name devices developed in his group for those in the lab as a way to encourage them to continue to work for him) into Barney Clark, a retired dentist. Clark was too sick to be eligible for a heart transplant, and the Jarvik-7 had just been approved by the FDA for human implantation. Barney's new heart had two ventricles and six valves made of titanium. The ventricles squeezed the blood, which passed through the titanium valves and out into the body, then returned through the other valves. It was powered by a pneumatic pump attached to a tube that trailed out of the heart (and out of the patient) like a sort of tail. Outside Clark's body, the pump was the size of a washing machine. One hundred and twelve days passed with Clark relying on this heart—miraculous days in the history of medicine. The *New York Times* heralded the success and suggested a fully implantable heart might be further off than originally expected but not more than ten years (which would have been 1994).[18] It was not, however, miraculous from Clark's perspective. At first Clark was delighted with his new heart, but then the problems began. Infections, a persistent issue with artificial hearts and assist devices, attacked his body chronically. His blood clotted in the machine. He suffered from strokes. He was not conscious for many of the 112 days, but when he was, he asked to die.

Only modest improvement has come since the implantation of Clark's heart. After Clark, a Jarvik-7 heart was implanted in another patient, Bill Schroeder, who lived longer, 620 days, long enough to take a call from President Ronald Reagan. But the extension of Schroeder's life, like that of Clark, was temporary and medicalized, not the great hope. Several companies now market artificial hearts. Versions of the Jarvik-7 have kept patients alive for months and, in a small handful of cases, a few years, but those years are rough. For a while, the Jarvik-7 was used as an "investigational device," which meant that it could be installed only in extreme cases in which there was truly no hope. And even then, its continued use depended on progress in its improvement. Then, in 1991, the Jarvik-7 lost its investigational status (too many years had passed without progress). New trials were halted until the device was able to renew its investigational status with a new name, CardioWest. New versions of these artificial hearts have electrical lines that run to large batteries in a backpack. This backpack of batteries, like the ones for assist devices, must be recharged every two hours. Ventricular-assist devices have become a relatively common means to help support hearts until they can recover (if they are able to) or as stopgaps until a heart-transplant donor can be located.[19] True artificial hearts are used when the heart requires more than assistance, when the living heart has nothing left to give. While this extra time is helpful (an artificial heart with a tiny internal component has even been used successfully in a baby in Italy), it does nothing to resolve the shortage of hearts. If there is a lesson here, it is that it is harder to produce a mechanical heart than it is to land a man on the moon or build a guided missile or a subsonic bomber, far harder. What evolution does with cells and mitochondria, we cannot yet do with metal, plastic, and batteries. This is a sentiment that has become most clear in the land of Salvador Dalí in an old church at the edge of Barcelona.

In Catalonia, the quixotic land of Salvador Dalí and Antoni Gaudí, a computational physicist named Mariano Vázquez has decided to build an artificial heart. Whereas DeBakey and Seaborg imagined devices that replicated the function of the heart (*what* the heart does), Vázquez seeks to model *how* the heart does what it does. His artificial heart will never be placed inside a body. It is a computer heart, a simulation. The idea came to Vázquez while having food and beers with a friend. Until that point, Vázquez had focused on engineering challenges—for example, how to build a better toilet or rocket. But his friend, in the way that friends will, asked him big questions. Why not focus on something more beautiful, more challenging, more interesting? Why not study the human body? If you can do anything with your life, why not try to make a working replica of, say, the heart? The heart is beautiful. The heart is mysterious. The heart, unlike a toilet or a rocket, has been shaped by millions of years of evolution. Vázquez, an Argentinean, had seen Domingo Liotta on television when he was growing up. He knew at least the rough outlines of the story of attempts to make an artificial heart. He would construct a simulation of a working heart instead. Vázquez and his colleagues at the Barcelona Supercomputing Center's project Alya Red decided to build this simulacrum by mimicking the ways in which individual muscle cells signal to other muscle cells in order to move the heart's liquid, its blood (much as another project mimicked the ways in which the flushing of a toilet moved other, less noble, liquids).

It is worth noting a beating heart does more than beat. It also rises to challenges. If a tiger chases you, a number of things will happen in your body. The amygdala in your brain will trigger a signal telling your body to run. That signal will travel to the adrenal glands, where the adrenal medullas release adrenaline. The adrenaline then makes its way to your heart's internal, biological pacemaker and speeds things up. It also makes the heart's beats more forceful. The adrenaline causes the cells of the heart to allow more calcium in, which makes more of the cells of the heart contract. The heart contracts both more often and more fully. If you find yourself running from a tiger, you will be very grateful for these responses of your heart.

This rapid reaction to a tiger is not the only response your cardiovascular system can carry out. Your heart also has sensors that allow it to detect how much blood is being pumped out of the ventricles. If too little blood seems to be in the system, these sensors trigger the production of more blood. They also trigger contraction in the arterioles of all but the most central organs (heart, brain, lungs). The heart is capable, in other words, of letting your fingers get a little cold to keep the heart, lungs, and brains from dying.

Vázquez and his team knew all the ways a beating heart did more than beat, and, in light of those complexities, they decided not to worry about any of them; they just wanted to replicate the details of an ordinary, resting heartbeat. To simplify things further, they decided, at least initially, not to worry about the movement of blood either. They would just simulate an unexcited, beating, bloodless heart. (It was probably a good decision; the Wright brothers, after all, did not make their first flight during a thunderstorm.) To do this, though, Vázquez and his team needed to know exactly how signals moved through hearts, and this required them to use a new tool to image the heart. Collaborating with scholars at the Computer Vision Center at the Autonomous University of Barcelona, they took high-resolution MRI scans of living hearts at a resolution of thirty-six micrometers, the width of ten

An example of a heart used in the Alya Red artificial heart models. Each noodle-like line is a muscle fiber based on imaging of a real heart. The Alya Red model is the best "total artificial heart" in the world, albeit one that is entirely virtual. *(Courtesy of Mariano Vázquez & Guillermo Marín, Barcelona Supercomputing Center)*

red blood cells. These images allowed the team to input data into their models on the paths of the muscle fibers in the heart, paths they turned into a kind of digital skeleton. They then modeled the ways in which an electrical impulse would travel along hundreds of thousands of simulated heart fibers overlaid on that high-resolution skeleton. The model specified the rules by which each of those fibers would contract when stimulated by adjacent fibers.

With the details of the high-resolution images, the simulation produced a working heart out of nothing but the high-resolution skeleton of a real heart and a few rules for the behavior of the muscle fibers. The model contains no master instructions, just directions for the behaviors of the individual fibers, but that is enough, much as the simple behaviors of individual ants can yield the sophistication of the colony. What is more, the approach is sufficiently flexible that, with high-resolution images from dog and rabbit hearts, the team has also been able to simulate the beats of the hearts of those animals. Just as in real hearts, each heartbeat is slightly different from its predecessor, a function of the specific sequence in the contractions of millions of cells. And each unique heartbeat of these simulated hearts is an indication that the scholars have come to understand the most important basics of how the arrangement of heart muscles and their signaling yields a beating heart. What was also unintentionally indicated, though, was the humility of scientists, engineers, physicists, and physicians. The computers that produced (and produce) these virtual, bloodless, unexcited hearts are so large, they require eight separate rooms filled with ten thousand processors. Here, then, is the real miracle, evidence of both our limited understanding and the heart's greatness. Those who attempt to create or improve artificial hearts must contend with the physical moral offered up by the supercomputer heart—namely, that the real heart is so complex that the best model of it we can make takes rooms and rooms, and even then, the model is far humbler than the real organ.

Vázquez and the team of more than thirty scholars with whom he works hope one day to stimulate not just an "average" heart but also particular hearts, yours or mine, so that the problems in individual hearts might be better understood and treated. Vázquez plans to add in a model of the flow of blood, maybe even the dynamics of the heart's responsiveness, though not quite yet.

Decades before Vázquez began building his computer heart, physician and essayist Lewis Thomas, describing the attempts to produce artificial hearts, wrote, "Not knowing the why of heart disease, we provided a makeshift device." But if Vázquez and his colleagues are able to simulate individual hearts, healthy ones and sick ones, they might one day also be able to use these makeshift digital devices to understand some of the electrical failures of hearts, some of the whys. They might even be able to model particular problems, such as my mother's arrhythmia, to understand where and how to treat the muscle fibers. At least in theory, they could model the clogging of the coronary arteries and the consequences. But for that, they would need to understand the blood, and even then, this approach would answer only one kind of why, the why of details and mechanisms. The other more difficult why is why these problems occur in the first place, why and when. A partial answer has been waiting for years, far from the supercomputers and surgical theaters, inside the body of an Egyptian queen.

9

Lighter than a Feather

*If thou examinist a man for illness in his cardia, and he
has pains in his arms, in his breasts and on one side of
his cardia ... it is death threatening him.*

— Ebers Papyrus

I first saw her in a photo. In the image, she looks forward. She is beautiful beyond reason. Her arms and chest are dressed in a honeycomb. She is holding two hard-to-distinguish objects, and her thin, dark neck rises proudly to her smooth face. The long braids of a wig cover her ears. She does not quite smile, but nearly. This is how she is remembered, in this single image, at age forty-five.[1] Then, of course, there is her body, which has come to change our understanding of the story of our hearts.

Born approximately thirty-five hundred years ago, she was the eldest daughter of Queen Nefertari and Ramesses II. She lived in the Valley of the Queens. She traveled with her royal parents as they were feted throughout Egypt, being feted herself in the process. At the dedication of the Abu Simbel rock temples in southern Egypt, she was honored with her own statue. She would have been carried around and tended to. Her diet would not have been a typical Egyptian diet, rich in vegetables and deficient in meat; it would have been a privileged one. She lived a good life. She may have eaten bread, olive oil, goat meat, pig meat, and honey, and she may have drunk beer.[2] Repeat. Add some delicious grapes. It went on like this, and then her mother, Nefertari, died. Her father remarried, and then the second wife died. Then her father died; when he did, Meryet-Amun became queen. She was a young queen for a very short period, and then she disappeared. This was, for a long time, all that was known of her fate.

Whatever happened at the end of her reign, when Meryet-Amun herself eventually died, we can assume her regal family spared no expense in readying and preserving her for what would come next. Specialists were called in. They cut open her small body beginning at her breastbone and lifted her organs out. All were tended to individually before being placed in ceramic jars, which would sit alongside her in case they might later be of use. Her heart alone was replaced in her body, which was now just a vessel, her posthumous boat. She was then wrapped up and sealed in a coffin on which her likeness was painted, and this was sealed inside another coffin, which in turn was sealed inside a giant sarcophagus. The sarcophagus was brought into her tomb. Inside her sarcophagus, her heart, buried beneath flesh and linen, continued to hide a secret, a sort of buried treasure that has recently been uncovered.

Her resurrection began in Egypt in 1940. Herbert Winlock, an explorer from the Metropolitan Museum of Art, was searching the Egyptian sands for statues of Queen Hatshepsut, the woman who had ruled Egypt as a king from 1479 to 1458 BC. Winlock was not alone. He was surrounded by a great army of men out on the sandy plains of Thebes. Each wielded a shovel or other instrument, all looking at the piles of sand for a hint of what might lie beneath.

Little was known at the time about their quarry, Hatshepsut, because her successor, a bitter son, all but obliterated the evidence of her reign, hacking apart each of the many statues of her likenesses, one by one.[3] Winlock wanted to find statues that might have been missed, though he knew as well as anyone that sometimes what you find is altogether different from what you were looking for.

Winlock's men worked an entire season without success. They

found ruins and bodies, of course—Egypt is full of both—but nothing of great significance. Then, one of the assistants saw something in the drifting sand,[4] up on the hills around the site where they had been working: pieces of shale stones where none seemed to belong. The shale might have been nothing at all. But, at least to Winlock's prepared and perhaps overly hopeful eyes, it looked like something. Could it be a pile of rocks tossed up where a tunnel had been dug, a tunnel into a tomb? He hoped. Winlock had been out in the desert for a long time, and any hint of discovery was enough to excite him. His heart pounded; it leaped at his throat.[5] He gathered his team and quickly they began to dig. They were trying to be careful, but from a distance, they looked like a pack of eager dogs, all searching for the same bone. Below them was the Valley of the Kings, and just beyond, the agricultural land along the Nile, rising like a green mirage. They dug and dug and then dug some more. They dug for a whole day and then forty-eight more, without any sign of discovery, all on Winlock's gut sense that somewhere beneath them was something more. Then, on February 23, the sand began to give way.

The foreman, Reis Gilan, reported to Winlock that the men had found a tunnel! It was rounded, shaped in the way that every tunnel—be it a mineshaft or an artery—seems to be shaped. The tunnel continued on to a brick wall. The wall looked as though it had been built hastily out of makeshift bricks. It was left unfinished and ready to be reopened. The men wanted to break the wall; they were dying to see the other side. It would be so easy! Like Winlock, they wanted to find treasure, whether for the excitement of discovery or because they had their own ideas about what they would do with it when it was found.

Winlock told the men to wait. If there really was an important discovery behind the wall, he didn't want everyone to know yet. He sent everyone home and guarded the hole. Some resolution, the men hoped, might come the next day (little did they know that the full revelation would take more than eighty years).

Several days later, on February 28, Winlock and a much smaller crew returned to the brick wall. They began to push carefully through. They hoped that, on the other side, they would find treasures—statues, gold, art, perhaps even a royal tomb (though over the weeks, Winlock's confidence had waned). Instead, they found garbage. There were scraps of things, bits of pots, baskets, and who knew what else. It was ancient garbage; in Winlock's words, "a rather disreputable rubbish hole." But the garbage was not all that had been left. There was a body, abandoned next to a small coffin. This body would turn out to be a small story, not what they were after. Farther in, they found a well, and beyond that, they could make out, by the light of their primitive flashlights, more of the tunnel. It continued, it seemed, on the other side and maybe, just maybe, there was a hint of a chamber, an atriumlike opening. Winlock went back out and told the men to go home. He stationed a guard at the site that, after so long in the desert without success, was beginning to seem, possibly, auspicious.

In the next days, Winlock had the men gather some planks, but before they could bring them into the tunnel, everything that had already been found had to be photographed and cataloged. Archaeology is a study in tedious anticlimax. Most days, it can be downright awful, the antithesis of what archaeologists new to the field hope it will be. Then, if you are lucky or prepared or both, there are the days when you get to jump over the well and see what is on the other side. Winlock knew that day was coming, but he was also beginning to sense that what he and his men had found was significant enough that he would be judged not just by his peers, but also by history; it would pay to wait.

Slowly, photos were taken; drawings were made. Every piece of garbage was labeled. It was not until the morning of March 11, a full two weeks after the tunnel had been found and nearly eight weeks after the men had started digging, that the well could be crossed. Finally, the wood planks that had been cut to be just the right size so as to fit down the narrow tunnel and around a corner and then over the well were taken in. It was not going to be easy. The men put one plank down and then extended the other. The planks not only fit, they seemed to settle into ancient grooves that must have been used thousands of years before to span the same gulf. Between the two planks, they set up a little platform over which Winlock could scurry. Winlock moved across and hesitantly started down the tunnel. As he did, his heart lightened, buoyed by possibility. His body was, in his words, "tingling with curiosity."

Beyond the well, the tunnel led into a large room where Winlock and his men could stand. In the room there were two enormous coffins, and inside the innermost coffin, a body. The body was wrapped in clean bandages, which were labeled, written on with detail and care. Here was a discovery. Here, also, was a series of mysteries, as time would prove.

Winlock was concerned with the mysteries of the woman's identity and the story of her life, and those would be partly resolved. Before Winlock left the coffin to the museum, he had gathered a few observations about the previous life of the body inside. Her chamber and coffins had been prepared with great care and expense, as had the small piece of cloth recovered from the well bearing her name; it was, she was, Queen Meryet-Amun.[6] Here was a real queen, the queen who had disappeared. She would have moved among the masses with special treatment, ushered on through her fortunate life until she died, whereupon she was ushered through her death. At least, that is, until Winlock got hold of her and then abandoned her in the back room of a museum.

The biggest mystery has turned out to be why and how she died and disappeared. She was not young for her time, and yet she died of something—some specific cause. Winlock couldn't tell, nor did he particularly care. He moved on to other mysteries in his own field of study, and the body of this queen was, in essence, abandoned. It was taken unceremoniously to the Cairo museum, where for ninety years it would simply wait. Her body had already waited since 1580 BC; what were a few more years?

The clues to Queen Meryet-Amun's death would eventually be found in her body, but her sarcophagus also offered a clue. There, Winlock found the normal elaborate scribbling one expects on Egyptian things: the picture-words of hieroglyphs and then literal pictures. One of the latter showed a heart being weighed on a balance against a feather.

When an Egyptian king died and went to the underworld, it was believed, his heart was put on a scale to see whether it weighed less than a single bird plume.[7] A light heart allowed passage into the afterlife, where a king could eat and copulate to his heart's content. The actions of a life were, it was said, scribbled on the heart, such that its weight was a fair measure of the king's actions. The idea that the measure of a man was written on his heart would reemerge in Christianity, in which it came to be believed that the heart contained a written record of sins and vice. The Egyptians were first, though, at least when it came to literally measuring the heart to judge a person's life. With time, this judgment was extended beyond the king to courtiers, the nobility, and even the priesthood.

In the hieroglyphs, a god was said to do the weighing, but the balance was watched over by a baboon that played the role of referee for admission to the afterlife, keeping his eyes on the fairness of the judgment. If the heart was light, the deceased could travel to the afterlife. Most Egyptian research on the body was related to ensuring a good afterlife rather than a long life. The daily life was the preparation; the afterlife was the real deal. Unless, of course, the heart was heavy, in which case both it and the chance for rebirth were eaten by the devouring beast Ammit, a Chimera with the head and jaws of a crocodile, the body of a lion, and the hindquarters and tail of a hippopotamus. It is tempting in retrospect to think of Ammit as a heart attack, the consequences of a heavy heart.[8]

It is not known whether any hearts were actually weighed or, if they were, to what sort of feather they were compared. It would have been in everyone's best interests to choose a heavy one, perhaps a lead plume. Many sarcophagi show images of the weighing of hearts, but among the most often discussed of such images is one found on Queen Meryet-Amun's sarcophagus. It is particularly elegant and clear. In it, the side of the balance holding the queen's heart appears slightly higher than the side holding the feather, but only slightly. The Egyptians, in other words, recorded the queen as having died with a heart just light enough to avoid Ammit's jaws. But Dr. Gregory Thomas at the University of California, Irvine, and his colleagues were about to prove that this was wishful thinking.

In 2008, Gregory Thomas visited Egypt for a cardiology meeting and, while there, toured the Egyptian National Museum of Antiquities in Cairo with his Egyptian colleague Adel Allam, a cardiologist who specialized in imaging hearts. In the museum, the two happened upon a mummy, that of King Menephtah (son of Ramesses II, born in about 1200 BC), which was labeled curiously. The text inside the glass case noted that Menephtah suffered from atherosclerosis. Atherosclerosis was, as Thomas, Allam, and all of their colleagues learned in medical school, a modern disease.[9] Something was clearly wrong with the label, or at least that was the first conclusion to which the two men came.

Thomas and Allam both work daily with the modern reality of atherosclerosis. They face a plague. In 2010, in the United States alone, seventeen million people died of cardiovascular diseases — more people than now live in New York City — and most of those deaths were ultimately the consequences of atherosclerosis. By all estimates, the number of such deaths is predicted to increase in the United States in the future, if for no other reason than because of population growth. The problem is not unique to the United States, though; the fates of Egypt's modern hearts are similar. As nations become developed, they are saved from the contagions and diseases of infancy and youth, and cardiovascular disease kills in their place. It is in this light that cardiovascular disease caused by atherosclerosis seems like a modern problem, one associated with Western life and diet. This is what we have been taught. This is what most doctors believe. Yet the truth is more complicated.

By the time Thomas and Allam saw the mummy in the National Museum, it was already known that heart disease is typically preceded by the formation of plaques in arteries. Those plaques are caused by inflammation that arises when cholesterol and certain immune-system cells (macrophages) begin to build up below the endothelial lining of large arteries. The word *cholesterol* is derived from the Greek *khole*, for "bile," and *sterol*, for "hard" or "stiff."[10] It is waxy and fatlike in appearance, but it is not a fat. It is a complex form of alcohol—a sterol—with the molecular form $C_{27}H_{46}OH$. Like all alcohols, its construction requires nothing but carbon, hydrogen, and oxygen. Cholesterol is necessary in your body. Without it, things go tragically wrong. But things can go tragically wrong with it too. Under certain conditions, the body reacts negatively to the cholesterol in the blood. Instead of simply being moved from one part of the body to another, the cholesterol is attacked by the immune system's bullies, macrophages and inflammatory cells. While we tend to think of plaques as fatty—yellow with cholesterol and other lipids—most of their mass is composed of the cells produced by our own immune cells. If our immune systems did not attack the cholesterol in the first place, the plaques would never form.

Why all of this happens has been poorly studied, but so has the simpler question of when, historically, it began. Compared to the history of heart surgery, heart transplants, and artificial hearts, the history of heart disease is not sexy; context rarely is. Christiaan Barnard, the man who first transplanted a heart, did not study why hearts become diseased or why hearts are sometimes malformed. For most of his career, he was not interested. Nor were his mentors. Nor were his competitors racing to transplant the first heart. They did not, except in passing, seem to really care about the history of the diseases they treated. For the most part, they did not even seem to have very informed opinions about their causes. Although they spent more time inside hearts than anyone else who had come before, these doctors were the mechanics at the accident—happy to have something to fix but not very aware of the tire tracks or other clues on the road. Of course, they saw evidence in details and problems, but they were too close to see anything resembling a big picture.

So why were hearts becoming diseased? Fate. Gods. Bad living. Bad luck. These were the things surgeons said when asked. They also might have said that heart disease was a modern condition. This assertion was based on intuition, but very, very few data. When Thomas and Allam became interested in the question of when hearts first began to break, the first discovery was that no one knew for sure when atherosclerosis had begun.[11]

Many diseases, most of them far less common than heart disease, have had their histories explored in detail. But those diseases tend to be the ones caused by pathogens. If a disease is caused by a pathogen, the genes of the pathogen can be studied to reconstruct its history. One can study the malaria parasite—plasmodium—or even the mosquito that carries the parasite from body to body and learn about the history of the disease that they, together, cause. But heart disease is not caused by pathogens (at least, not directly). No set of genes exists to be considered other than the afflicted host's genes, whose stories are often hard to translate. To understand the origin of heart disease, then, one of the only recourses is to look for ancient examples of hearts, and for this, one needs bodies.

Despite having no experience whatsoever in working with ancient bodies, Thomas and Allam found themselves compelled to study them. During the reign of Napoleon, the Rosetta stone was discovered in Egypt and allowed the messages in the hieroglyphs to be read. Thomas and Allam thought the bodies of ancient Egyptians might allow scholars to decode another mystery, the story of our weak hearts. For one, there were many ancient Egyptian bodies in museums. For another, ancient Egyptian society was a reasonable cultural antecedent of modern society. (The influence is direct; the culture of Egypt strongly influenced the culture of Greece, and the culture of ancient Greece has influenced every bit of modern Western culture.) Thomas and Allam were experts in the heart, and Allam in particular was an expert on imaging the heart and blood vessels through the skin using CT scans. It could not be so much harder to do the same through the mummies' wraps.

Soon, Thomas and Allam had enlisted a large number of colleagues to work with them, colleagues who were experts in each of the necessary steps and details of the process (a photo of the team in 2010 shows nineteen smiling scholars).[12] But in order to compare modern hearts to ancient ones, Thomas, Allam, and their team needed mummies. But working with Egyptian bodies is now an ordeal nearly as elaborate as becoming mummified, and for good reason. The mummies are not just artifacts; they are also ancestors, and therefore, handling the mummies requires layers of approval so enveloping that if one is approved, it feels as though passage has been granted to the afterlife (it often seems far more likely that the research proposal will be consumed by Ammit, the crocodile beast). All Winlock had had to do was buy some shovels and start to dig. But for Thomas and Allam, the approval process consumed valuable time. Finally, amazingly, after testifying before a committee of seventy Egyptian archaeologists, writing and more writing, and then, when all of that did not seem to be working, a bit of begging, Thomas's application was approved. It helped that his study seemed to be important. Also, most of the individuals on the panel determining whether he would be approved were of an age at which heart disease would have concerned them in more than just abstract ways.

The team came together in Cairo in 2009, where they were taken to the Egyptian National Museum of Antiquities in Cairo. At the museum, they were asked to choose the bodies they wanted to study. They chose the 45 best mummies out of the 120 or so available. Inevitably, the team knew less than was ideal about any particular body's history. They were researchers, clinicians, and surgeons, not archaeologists. Yet they were so tickled by the possibility of studying mummies (who wouldn't be?) that any body would and could work. The queen's was among the chosen, but the doctors did not yet know her story or anything about Winlock. Hers was just another body among the many stacked in the museum in the most ordinary of ways.

Once the bodies were chosen, the men carried them to the hospital, where they would be reweighed, at least figuratively. The men looked like pallbearers, but they were on their way to a rebirth. The coffins were heavy. But coffin after coffin made it down the road and up the elevator to the whole-body CT scanner, where the bodies (including the hearts, if they were present) were slid into the machine just as you might be if you were getting a CT scan. Not surprisingly, a richness of details was visible through the shrouds of cloth wraps and skin. Even through the wrappings of mummification and time, atherosclerosis showed up on a CT scan, a white smudge where a plaque has calcified. With the scans, the hearts were weighed once more, though not against a feather this time. They were weighed against modern hearts. Here would be the test of just how much the hearts in the good old days were like the hearts in the bad new ones. The scientists compared the images they made of the ancient hearts to images of modern hearts to see if the modern condition really was modern or, instead, a reflection of a deeper human condition.

Some of the bodies were from individuals who'd died relatively old, others younger, though they were all adults. Some, such as Lady Rai (who died in about 1530 BC), nursemaid to Queen Ahmose Nefertari, were from the first days of the Egyptian empire, when only kings and those closest to them were mummified. Others were from later days, when far more individuals had a shot at the afterlife. The earliest was from 1981 BC, the most recent AD 364. The bodies were those of people who had not previously interacted with one another. Using them to characterize "ancient Egyptians" is like using a dozen people you meet at the queen's palace in London to characterize modern Europeans, and yet it offered the best look to date at the ancient history of our hearts. None of the bodies would have been from working folks. While mummification became increasingly common as time went on, most Egyptians and all of the block cutters and builders, for example, were buried with little fanfare in the drifting sand.

Though it was possible, the scientists and doctors did not think it was likely that they would see concrete evidence of the ultimate consequences of atherosclerosis: heart attacks or strokes. What they could find, though, might be the white reflection of the atherosclerosis itself. In modern bodies, the presence of atherosclerosis in any part of the body is a good predictor of its presence in the arteries of the brain or even, that most dangerous place, the arteries of the heart. It was also possible the men would find no atherosclerosis at all, even if it was there.

If the ancient Egyptians were found to have atherosclerosis, it might be due to their similarities to modern humans. Although removed greatly from us in time, wealthy ancient Egyptians may have eaten and lived enough like modern Egyptians or Americans to have similarly plagued hearts. The Egyptians were, in general, thin. As Allam and colleagues put it in one article, "After all, the ancient Egyptians built the great pyramids by hand without modern machinery, and 'fast food' was not yet invented." But the individuals prestigious enough to be mummified were also, almost certainly, the same individuals likely to eat a lot and move relatively little. The food available to the most gluttonous Egyptians, like that available to modern Americans, included meat, dairy products, eggs, processed grains, salted meats, and beer. Researchers studying Queen Hatshepsut have described her as a big woman with "huge pendulous breasts."

But there was another, more interesting possibility. Perhaps atherosclerosis is not associated only with rich, slothful living. Perhaps it is more universal than previously suspected. The Egyptian data would not, on their own, allow Thomas, Allam, and crew to distinguish between the good-living hypothesis and the universal-atherosclerosis hypothesis, but it would be a start. Despite what the sign on the mummy in the museum had said about Menephtah, the most likely scenario by far was that the mummies would be devoid of atherosclerosis.

The team examined the mummies as a group, looking at CT scans together and conferring, the way doctors might about a living patient. They consulted with one another to reach a consensus about what each scan showed. They looked together and what they found astonished them. Vascular tissue, be it from veins or arteries, could be identified in forty-three of the mummies, and at least pieces of hearts could be found in thirty-one. The doctors saw plaques in one body, and then another, and then another. Amazingly, plaques were visible in nearly half (45 percent) of the mummies. The plaques were more likely in older mummies, just as they are more likely in older people today,[13] but were present in mummies of many periods and across a range of social positions (at any rate, from pharaoh to pharaoh's attendants).

In general, the mummies provided clear evidence that cardiovascular diseases were likely to have been common among Egyptians, at least those Egyptians who were mummified, even if the individuals did not often live long enough to die of those diseases—though some of the mummies may well have died of their blockages. One of the mummies was the husband of Queen Hatshepsut. It was when he died—perhaps as a consequence of atherosclerosis—that Hatshepsut became queen (their oldest heir was too young to rule). But the team did not initially see in the mummies what one might expect to find in a modern sample of people of ages similar to those of the mummies: atherosclerosis of the coronary arteries.

True, the hearts were present in only a minority of bodies. But still, it would have been nice to confirm that atherosclerosis of the coronary arteries had occurred at least sometimes. Seeing a clogged coronary artery would be as close as one could hope to get to evidence of an ancient heart attack.

The team members found only a few coronary arteries they could easily study. But then they looked at Queen Meryet-Amun after sliding her into the CT scan. The queen had atherosclerosis in every artery bed examined, including her coronary arteries. Her arteries looked like those of someone who might turn up at the hospital having just suffered a heart attack. Thomas would later say of her condition that if she had come into his office, he would have recommended a double bypass. Her body, although draped in ancient finery, could have passed as modern. The Egyptians, Meryet-Amun's body revealed, were not just a little like us; their hearts and our hearts seemed the same. Meryet-Amun might even have died of her broken heart; that might be why she disappeared from power in the first place. If we are to reconsider the drawing on her sarcophagus, her heart was, in a way, far heavier than a feather. It was weighed down with our "modern plague." Nor are the Egyptian bodies totally unique. A later study found that the Tyrolean iceman discovered in present-day Italy, a man who'd died from an accidental fall at the age of about forty-five around 5,300 years ago, had plaque in his carotid arteries, as well as in his aorta and iliac artery.[14] Clogged arteries seems to be an ancient human phenomenon and yet there was still a lingering doubt in the minds of Thomas, Allam, and others. They had proved atherosclerosis to be ancient, but they could not distinguish whether it was ancient because humans have, for many millennia, indulged in lifestyles that cause their arteries to clog or because atherosclerosis is universal, a condition of being human. Maybe poor Egyptians (the men and women who built the pyramids, for example) and hunter-gatherers were very different. Perhaps atherosclerosis was only as old as excess.

Thomas, Allam, and friends thought they could figure the answer out. The Egyptians were not the only people to mummify their dead. Mummies can be found around the world in those places that are cold or dry enough that one can bury one's cousin and, without too much trouble, ensure that he or she will look the same in a few thousand years. In many places, such as Egypt and Peru, mummies are intentional, but in others, they just happen; for example, when the dead are left in cold, dry caves. And so, Thomas, Allam, and their colleagues reasoned, one could, in theory, consider the presence of atherosclerosis in mummies from around the world. One could also do this in practice. Thomas and Allam brought together an even bigger team than they'd worked with before to evaluate 137 bodies: mummies from the end of the Roman era, mummies from ancient Peru (900 BC to AD 1500), mummies of the ancient Pueblo of the southwestern United States (1500 BC to AD 1500), and five mummies of the hunter-gatherer Unangan people from the Aleutian Islands (about AD 1800). Together with the samples from Egypt, they provided bodies from farmers and hunters-gatherers from different periods, cultures, and regions.

Here I will pause for a personal aside. When I started writing this book, the results from this broader study were not yet published. Knowing what I know about hunter-gatherers, farmers, diet, atherosclerosis, and the heart, I would have predicted that the hunter-gatherers would be free of atherosclerosis, as might be many of the less affluent individuals from agricultural cultures outside Egypt. The folks in Peru, for example, had diets based heavily on fish. The Pueblo ate farmed foods, but far different kinds than the Egyptians. In light of how extremely different the lives of these peoples were, I would have bet quite a bit of money that they would vary greatly in their atherosclerosis. It is clear from interviews done before this study of members of the team that many of them would have bet the same as I did.

And we would all have lost. Just as in modern humans in affluent cultures, the degree of atherosclerosis tended to increase with age. Atherosclerosis was found in every sample of mummies, from each region and time. Probable atherosclerosis was discovered in 34 percent of the mummies. It is not modern; it is our ancient condition. Sixty percent of the Inuit and Unangan hunter-gatherers suffered atherosclerosis, which is to say that hunter-gatherers were in worse shape than the agriculturists, exactly the opposite of what everyone would have predicted. These results turn on its head the idea that atherosclerosis is a modern condition, the result of our terrible lifestyles. Our terrible lifestyles play a role, but, as the authors of the study put it, "The presence of atherosclerosis in pre-modern human beings suggests that the disease is an inherent component of human ageing and not associated with any specific diet or lifestyle."[15]

Had the surgeons working in the 1960s and 1970s known all of this, and had they had the perspective of time, they might have paused to reflect upon what was going on in the heart. They might have begun to think of coronary atherosclerosis in particular as a natural death, a part of aging we want to forestall and yet, still, part of aging. This is what da Vinci thought. Or they might have approached it from the perspective of evolution to understand whether coronary artery disease is even more ancient than ancient Egypt, Peru, and North America. Instead, upon confronting the atherosclerosis that Sones and others saw early on in coronary and other arteries, those who weren't trying to transplant hearts or build artificial ones began upon a trajectory suggested by technology, a trajectory with an emphasis on intervention, a trajectory toward precisely the technique, coronary bypass surgery, that Gregory Thomas suggested Queen Meryet-Amun needed for her heart.

10

Mending the Broken Heart

René Favaloro's story began as a classic tale of self-motivated accomplishment. Favaloro was born in 1923 in a working-class neighborhood of La Plata, a town thirty miles south of Buenos Aires, to Sicilian immigrant parents. His mother was a seamstress, his father an artful carpenter. While Favaloro lived at home, he stood with his father and built useful furniture. He could have followed in his father's footsteps. But for some reason, even in grade school, he imagined himself as a doctor, and so he kept moving, going to college and then, eventually, medical school (although each year he would come back on his breaks to help with the details of carving and veneer work). When he graduated in 1949 with his medical degree, his mentors agreed that he was a great young doctor. But there was one thing at which he was truly outstanding: working with his hands. He had the hands of his parents, with a carpenter's strength and a seamstress's subtlety.[1] He trained himself to work with both his dominant right hand and his lagging left. He would hold his breath and cut and sew with both hands. When he did, he could weave together the body so perfectly that it seemed as though it had never been apart. Everything was going right. Favaloro was headed for a successful life as a big-city surgeon, his high-school sweetheart, by then his wife, Maria Antonia, at his side. But this simple future was not to be.

In 1946, while Favaloro was finishing his medical training, Juan

Domingo Perón was elected president of Argentina. Perón quickly concentrated power, including power over the contracts of many doctors and academics, power he used to quell dissent. Faculty who disagreed with Perón, and there were at least fifteen hundred of them, resigned or were fired. Favaloro knew all this; he had watched his colleagues and mentors leave. The oppression these departures symbolized boiled under his skin. He was an idealist in a country where such people faced difficulties. This idealism had its first test when Favaloro graduated. He was offered a prestigious job, but on the condition that he profess his loyalty to the Perón's Justicialista Party. He refused. This refusal exiled him from big hospitals for as long as the Peróns were in power. This man who exuded a kind of inevitable greatness left for a small village in the southwest of the gray dry pampas in 1950. He found a house he could turn into a clinic. All around the wind blew. Cows stood in the grass. He was, as he would often say later in his life, an urban kid who had suddenly become a country doctor.

Favaloro did what he could and built the old house into a clinic. He figured out how to make operating rooms, a sort of lab, and even x-ray equipment. In his spare time, he helped care for his brother Juan Jose, who had lost his leg in a car accident. Two years later, his brother, also a doctor, was rehabilitated enough to join him in the clinic. Side by side, for twelve years, the two brothers worked with great success. They dramatically reduced the rate of infant mortality in the region. Meanwhile, Favaloro kept reading and learning about the rest of the world. The rest of the world was leaving Argentina behind in science, in medicine, in everything, especially surgery. Favaloro told his brother and his wife, over wine at the table in their village clinic, that what they were doing was medicine that was already accepted rather than the cutting-edge treatments yet to be discovered. Argentina deserved more! René Favaloro wanted his country to progress, but he could not lead his country out of its past while Perón was still in power. Much as it pained him, he would need to leave, to go somewhere. One of Favaloro's mentors, Professor José María Mainetti at the Universidad Nacional de La Plata, advised him to look to the Cleveland Clinic in Cleveland, Ohio, in the United States, where, he said, they were making progress. Mainetti told Favaloro that he would write his friend George Crile Jr. at the Cleveland Clinic to put in a good word.

As much as Favaloro was a successful surgeon and a hopeful idealist, he was not a practical man. Impractical men who succeed are called visionaries. Impractical men who fail are called failures. The jury was still out on Favaloro, and so, without any assurance from Cleveland that he would have a job there or even that anyone had received Mainetti's letter, he decided, at the age of forty, to uproot himself and his wife and go. He would go to the place where the greatest surgeries were happening. He would go there to do work and maybe someday bring his successes back to Argentina once the Peróns and their legacy were gone. He would do all of this, and so, in 1962, he bought tickets for himself and his wife.

When the two arrived in Cleveland, Ohio, Favaloro dropped his wife and luggage at the hotel and showed up at the office of the medical center's chief of surgery, George Crile. Favaloro was used to leading. He'd earned it. He'd built and run a successful clinic. But in Cleveland, he had to explain that he had just gotten off the plane from Argentina and that he would like a job. This would be presumptuous anywhere, but Favaloro was asking for a surgery job. He could start the next day, he said. What, he asked the chief, should he do first? Crile resisted laughing. He had never gotten a letter from Favaloro's mentor; this was the first he had heard of Favaloro. He politely sent Favaloro down the hall the way you might invite someone to take a long walk off a short pier.

Favaloro was sent to another door, and then another, until he finally made his way to the door of a man who could make sense of what was going on: Don Effler, a close colleague and frequent adversary of Sones. Effler was an accomplished heart surgeon and was fully aware of the layers of ridiculousness in what Favaloro was trying to do. Effler took the time to tell Favaloro what Crile could not be bothered to say, that one needed a U.S. medical degree to do surgery in the U.S. Favaloro appeared not to have known this (or anything else about the place he had landed). He went back to his hotel, where he joined his wife, defeated. What had he done?

The next day he went back to the clinic. He told Effler that he would "immediately start studying for the exam and...work for free" on whatever Effler would have him do. Amazingly, Effler consented to this plan and got Favaloro started doing the most menial of tasks. Favaloro worked hard, and from those tasks, he would rise, rapidly, earning his U.S. medical certification by night. Before anyone knew it, he was, once again, a surgeon, a surgeon with grand ideas, and the only man who was able to work with both Effler and Sones. With Sones, Favaloro studied coronary angiograms, thousands of them, hour after hour. With Effler, Favaloro dreamed up ways to mend the problems that he and Sones had seen.

When Favaloro was getting started in Cleveland, there were still no direct treatments for most heart diseases, including the most common set of problems, those observed by Sones, those due to the blockage of the coronary arteries. In the early 1900s, the main symptom of a temporarily blocked coronary artery, angina, had been treated by covering up the pain.[2] The nerves that conveyed the pain from the heart were removed in many thousands of patients in a treatment akin to pulling out wires in your car radio to hide the sounds of the static. Alternatively, the thyroid was destroyed, which slowed metabolism and blood flow and reduced the chance that the heart would clog, at least in theory.[3] Still other treatments called for irradiating the heart or injecting alcohol into the vertebral column. These methods were crude and yet still in practice in the 1960s even as bold surgeons had begun to contemplate heart transplants.

But some surgeons had the idea to graft or redirect wider, less constricted arteries from elsewhere in the body onto or into the heart, to replace or add to the function of the afflicted coronary artery. It was an idea akin to the heart transplant but with key advantages. This kind of surgery did not require a donor, so surgeons did not have to find a way to get past the response of the immune system to foreign tissue.

Transplants of arteries had been discussed for decades. In 1910, Alexis Carrel wrote that "an arterial wall can be patched with a piece of artery, or vein...these operations present very little danger, and their results, observed many months after the graft, were excellent."[4] Some had followed up, hesitantly, on Carrel's work. Several surgeons had tried to redirect arteries and veins so that they might spill blood into the area between the pericardium and the heart, thinking that it might be absorbed. In 1946, Dr. Arthur Vineberg at McGill University had another idea. He cut the mammary artery and redirected it to the heart, where he sewed it into the ventricle wall. One group reported on angiographies of eleven hundred cases in which patients had been treated with "the Vineberg method." In almost half the cases, blood flow to the heart muscle from the mammary artery appeared to have been achieved.[5] This was success, and yet, because the cases in which the procedure did not work were often fatalities, modest success.

Favaloro thought he could improve on these approaches by following up on Carrel's suggestion to transplant an artery or a vein rather than just redirect one. In 1954, Gordon Murray at the University of Toronto reported on successful transplants of three different arteries (subclavian, carotid, and mammary) into the hearts of dogs, with auspicious results.[6] Other dog surgeries had followed in other labs with similar positive results. After looking at the details of these studies and of the many angiograms Sones had done, Favaloro thought he was ready to try to transplant the saphenous vein (a large vein from the thigh) into a human heart.[7]

He knew he could sew almost anything. This was not a skill possessed by all; it was what made him special, and he needed to take advantage of this gift. Favaloro would cut a section of saphenous vein from a patient's thigh. He would then turn it into a bypass by sewing one end of it to the aorta and the other end to the coronary artery just beyond the area of the blockage. Other people had tried to do this, but they'd done it in emergencies rather than as part of a reliable solution to clogged coronary arteries. One attempt, in 1964, by three surgeons at Methodist Hospital in Texas was successful, but it was not reported until 1973.[8] Favaloro would do it with his own artful simplicity and, because he was working with Sones, with Sones's vision of a repeatable, common solution. If such a thing could be done, it would mean not only that this specific procedure — a bypass, as it would come to be called — was possible, but also that many other things were possible.

Favaloro talked to Sones about the procedure, and the two agreed that it should be done only on a patient whose right coronary artery was totally blocked in one section but open in others.[9] One day Favaloro found himself with a patient that he could not fix using any standard approaches. The patient, a fifty-one-year-old woman, had a nearly completely blocked right coronary artery. If sewn back up after having been investigated, she would almost certainly die. The vessel would become entirely blocked, her heart would stop, and the blood would fail to reach her brain. Favaloro thought he could do something using his magic hands, and so he hesitantly began.

Favaloro cut open the woman's chest. He did not yet have a heart-lung bypass machine (although one was present at the hospital) and so he would have to work quickly. He cut a section of saphenous vein out of the woman's groin and then cut out the woman's right coronary artery. The latter was every bit as clogged as he had imagined it would be, a thin tube filled with plaque. Then he began to sew, as only he could, like a seamstress of flesh: carefully, delicately, his small fingers pushing the tissue in and out, holding it just right. A poor stitch, and blood would pour out into the body. A missed stitch, and nothing would work. Favaloro did not breathe while he sewed; he held his breath so that his hands were in his control—he owned their motions. Then it was done. He exhaled. The blood was allowed to flow back into the heart. The woman was sewed shut and brought into a room, where a catheter was run up through one of her vessels and into her heart. Dye was released into her heart, and an x-ray was taken. What Favaloro hoped to see was a black space indicating that blood was moving from the heart through the new artery. None of his colleagues believed it would be there. The object of the quest was remote and improbable. There had been no way to know if it would work. If it didn't, this woman was about to die.

Then Favaloro and his colleagues turned on the x-ray and saw the blood moving through the heart and flowing through the artery. They saw it! Mason Sones, the de facto adviser to Favaloro, ran out into the hall and announced, "We have made medical history." He had yelled it before; he would yell it again. In that moment, for Favaloro, everything became worth it: the exile to the pampas, the trip to Cleveland, the time spent working as a helper for free, the long hours, all of it.

There would be improvements.[10] Favaloro soon realized he did not have to cut out the old artery from the heart. He could just leave it in place and run a new artery alongside it, as a bypass. This is what would give the surgery its modern name, coronary artery bypass graft, or CABG. He also tried the left coronary artery. He then grew even bolder. He performed the bypass during and even after a heart attack (an acute myocardial infarction). In December of 1968, just one year after the first surgery, Favaloro was able to publish a report of 171 patients on whom he had conducted bypass surgeries. Half of those patients had received double bypasses with mammary artery implants. Just two years later, 1,086 bypasses had been performed by Favaloro and others he inspired. To see this growth in the rate of bypasses, one might suspect Favaloro of too much ambition, but the mortality rate of his patients was just 4.2 percent.

Favaloro had achieved success most can only dream of. Favaloro, the son of an Argentinean seamstress, was even invited to come on talk shows along with Sones and Effler.[11] He was saving people nearly every day (or at least every day he wasn't on a talk show). This man from a poor neighborhood in La Plata was mending real, beating hearts, and yet this was not enough for him; he was dissatisfied. Four years after the first coronary artery bypass, Favaloro decided to move back to Argentina. He had, in a few years, brought greatness to the United States, and now he wanted to return to Argentina to bring greatness back to his own country. He would go on to start a Cleveland Clinic for Argentina.

Even as Favaloro was traveling to Argentina, another story was unfolding, one that would ultimately change the place of Favaloro's work in the story of the heart. More than that, it would plant the seed that would lead to the decline of heart surgery itself.

Adolph Bachman came into the Medical Policlinic Hospital in Zurich, Switzerland. He was near death, though he did not know it initially. The thirty-seven-year-old came in complaining of chest pain, angina. The doctors performed an angiogram on him to see his heart. A three-centimeter stretch of one of Bachman's coronary arteries was nearly entirely blocked. Most of the artery looked fine; the angiogram showed a black river of blood where one should be, but then that river narrowed so much that whatever stream was present was invisible and obscured by collapsing riverbanks of plaque. Had he understood the angiogram, Bachman might have been worried, but the doctors did not look upset, so he would be strong. In fact, they smiled over him, like a flock of vultures.

Initially, the doctors told Bachman he would require a bypass, Favaloro's new surgery—but at the last minute, an alternative possibility emerged. One of those doctors standing over Bachman was Andreas Gruentzig, who had invented something he really wanted to try out. A typical angiogram catheter has a narrow tip at its end through which dye is released. But Gruentzig had produced something unique after spending the last ten years experimenting and tinkering. At the end of his catheter, he had fastened a sort of sturdy balloon.

Gruentzig had tinkered anywhere he could, but the fortuitous last tweaks occurred in his apartment, where he and his assistant, Maria Schlumpf, did the work on his kitchen table. A picture still records the event. Scattered around them were the bits and pieces necessary for the endeavor—plastic, Krazy Glue, a bottle of wine, and balloons. A balloon on the end of a catheter seems like an invention one might see at a high-school science fair. It took thousands of tries to get it right, but the fundamental technology was ridiculously simple: a balloon would be inserted into the artery and inflated, the pressure of the inflation would make the narrow artery wider, the balloon would be removed, and more blood would flow through the vessel as a consequence. But it was one thing to test such a device in a kitchen and quite another to expand it inside the most intimate stretch of a man. Bachman would be that man; at least, he would be if Gruentzig could persuade Bachman to let him try the new invention.

Although Gruentzig had spent long hours working on his device and perfecting it, he had tested it during bypass surgeries only in arteries that did not need strong blood flow and would be either bypassed or clipped out. Bachman was the perfect candidate for a real attempt. Gruentzig explained the device to Bachman; he then told him something that would be said repeatedly by cardiologists to their patients: This approach will be much easier and will require less recovery time. We don't even have to open up your chest.

Bachman was convinced. He signed the necessary forms, and, almost before Bachman lifted his pen off the paper, Gruentzig had inserted a catheter into his right coronary artery. He then pushed it up until it arrived in the heart. Once there, Gruentzig inflated the balloon, and as he did, he held his breath. Everyone could see the balloon begin to push the artery outward. That the result would be a success was not obvious. The artery could rupture. Or it might snap back in place once the balloon was removed, just as narrow as it had been.

Miraculously, the procedure worked. The balloon expanded the artery, and blood flowed freely again. It was as simple as unclogging a drain. It was just the tools that took ingenuity. The patent for this procedure would quickly make Gruentzig millions of dollars and would lead to hundreds of other devices; once Gruentzig had shown that the arteries could be successfully manipulated, a small army of tricks and tools for the ends of catheters emerged. Among the most significant of these new tools was the stent, which was to be used following angioplasty. It was a small metal-mesh tube that could be left behind and used to hold the artery open more predictably once the balloon was extracted. It was a sort of permanent angioplasty. From Gruentzig's lead, an entirely new type of cardiologist has emerged, a specialist who never cuts open a heart but instead, like a spelunker, explores its caves.

In addition to its apparent success, angioplasty, which later included the placing of stents, had other things going for it. It made sense. The balloons and stents are biophysically intuitive medicine. You take a pipe that is clogged and push the clog out of the way. You take a pipe that won't stay open and you reinforce it. These were the sorts of solutions plumbers might use, and so it was easy for the procedures to catch on. They spread rapidly, eventually at the expense of bypasses. Coronary artery bypass surgery had become ever more common in the seventies. Then, on its heels, angioplasty and then stents became ever more common. All of this happened with very little consideration of what an optimal solution would look like. Initially, it happened without a single study comparing the fate of patients who underwent bypass surgery with that of patients who had stent treatment. Such comparisons would come, but only much later.

With time, Gruentzig's method has been elaborated upon. There are now many kinds of stents, some of which, in addition to holding open the artery, can release (*elute*, the doctors say) drugs.

In other words, without ever opening up the chest, doctors can widen an artery, reinforce it, and implant a device that releases drugs directly where the artery seems most blocked.

While Gruentzig's methods were proliferating, Favaloro was back in Argentina. He had decided to return just as his creativity as a researcher and doctor was at its maximum, at forty-seven years old. He could have stayed in the United States and found wealth and fame. It was a hard decision, but he had begun practicing medicine to help the people of Argentina, and he needed to return. He handed in his letter of resignation at the Cleveland Clinic and left a note on his boss's desk: "As you know, there is no real cardiovascular surgery in Buenos Aires.... Believe me, I would be the happiest fellow in the world if I could see in the coming years a new generation of Argentineans working in different centers all over the country able to solve the problems of the communities with high-quality medical knowledge and skill."

Back in Argentina, Favaloro began to take the steps to establish a major clinic. In 1980, he created the Favaloro Foundation (which would later become Favaloro University). There, he taught more than four hundred residents from Argentina and elsewhere in Latin America. They were trained in heart surgeries, like Favaloro's cardiac bypass, and in other techniques, including, eventually, angioplasty and stents, as well as in approaches for dealing with liver, kidney, and other organs' problems. Again, Favaloro found success — greatness, even. This humble, ambitious man was sewing hearts back together all over Latin America, and he was inspiring others who were mending even more. And then tragedy struck.

The economy of Argentina began to collapse in 1998, precipitated by the breakdown of the economies of Russia and Brazil. Suddenly, the Favaloro Foundation found itself seventy-five million dollars in debt. Everything Favaloro had fought for seemed to have stalled. Favaloro was depressed. His wife, who had been at his side through everything, had recently died. Favaloro was seventy-seven years old and had seen many ways of dying. Whatever the reckoning, he wrote a letter in which he laid out what he had accomplished and what he had failed to accomplish. He noted that he was sick of "being a beggar" on behalf of his people and their broken hearts. He then picked up his gun and shot himself in his own beating, bleeding heart.

Favaloro had gone home to Argentina to give back to his people, but Gruentzig had taken a different trajectory. On the back of his successes, he lived an ever more adventurous lifestyle and had no interest in returning to Germany. In 1980, he had been given U.S. citizenship on the grounds of his being a "national treasure." In his lifetime, more and more people had stents implanted. But Gruentzig would never need to undergo this procedure himself. On October 27, 1985, Gruentzig and his wife decided to fly from St. Simons Island to Atlanta in a rainstorm. Their plane crashed near Forsythe, Georgia. Gruentzig was just forty-six years old, and his wife, Margaret Anne Gruentzig, a medical resident, was just twenty-nine. The hearts of Gruentzig and Favaloro never faced the atherosclerosis the men fought so hard to remove.

But Bachman, the first angioplasty patient, lives on. In 2007, the coronary artery Gruentzig had worked on was checked by Bernhard Meier, the doctor who took over Bachman's care in Switzerland after Gruentzig moved to Atlanta. The artery was slightly more occluded than it had been but not enough to warrant treatment. Bachman finally quit smoking in 2007 and has tried to reduce the stress in his life. His doctors did not recommend a stent, but Bachman insisted, and so his wish was granted. As of the writing of this book, Bachman lives, as do the techniques of Favaloro and Gruentzig.

By 1970, coronary bypass had become one of the most common surgical procedures in the world. By 1990, angioplasty[12] (to be augmented by stents) had become even more common than bypass surgery. In some places, such as the United States, success would be measured purely in terms of medical success—the additional years of life someone lived, his or her health and well-being (in such places, both angioplasty and bypasses are common). However, success can also be measured by the time patients spend in the hospital and the money hospitals make. Medically, Gruentzig's stents seemed similar to bypass surgery in terms of their success, no better, and maybe they were worse, though the benefits seem to depend to a great extent on the condition of the patient. But stents required the patients to spend far less time in the hospital and made hospitals far more money, and so they quickly became the most prescribed treatment for blocked arteries. The fingers of surgeons continued to relocate veins from legs and groins and use them to reroute blood around blocked arteries, but fewer times each year. Heart surgeries in general became less common.

While stents and bypasses both let the river of blood flow freely again, they do not change what causes the atherosclerosis in the first place. Atherosclerosis is like a busy beaver. It finds the flow, wherever it might be, and stops it. I spent much of my childhood unclogging the culverts that drained a pond at my house. A beaver clogged the culverts with feces, spit, sticks, and mud, and it was my job to clean them out so that the water might flow. But as long as that beaver was there, the culverts would clog again.

The beavers in your body never stop when you have atherosclerosis. This is a problem neither Gruentzig nor Favaloro nor any of the other great minds of their surgical generation solved. They cleared out the debris, but bypasses could clog again. About 15 percent of bypasses clog in the first year, and by ten years, 40 percent have clogged.[13] Reopened arteries can clog again. These procedures were beautiful, miraculous, and mere stopgaps.

Someone would need to find and stop whatever it was that was creating the clogs in the first place.

11

War and Fungus

In 2004 Akira Endo went to his doctor. He was feeling fine but was due for his checkup. At the doctor's office, the doctor did the ordinary tests. He drew some blood, he checked Endo's pulse, he took notes—and then he sent him home and told him to wait. Endo was at an age when the body has a tendency to break down without notice. Endo felt well, but he might not have been.

Days later, the phone rang. It was the doctor with bad news. Endo's cholesterol level was too high, 240 mg/dl, with an LDL of 155.[1] There was also good news. Some "very good drugs," called statins, could be used to lower cholesterol.

Endo began to laugh.

Endo's prescription from the doctor was among the most universal of modern medicines. In the United States, one in ten adults is now taking a statin drug. One in three people over the age of sixty-five takes a statin. Some researchers have suggested that every adult should take statins, so useful are they perceived to be in preventing heart disease and strokes. Endo's prescription was ordinary. Less ordinary was Endo's longer relationship with these drugs and the fungi that originally produced them.

In July of 1945, when Endo was eleven, atomic bombs were dropped on Hiroshima and Nagasaki. Mushroom clouds rose above those two towns, evidence of the powers of science and of the horrors of man. The war ended, but not before it had taken hundreds of thousands of lives, along with Japan's prosperity. Endo lived on a farm in northeast Japan, far from Hiroshima and Nagasaki but well within the reach of the war's economic consequences. After the bombs, the country was occupied, and food grew scarce, so Endo, like many, honed his expertise in gathering, both for his family and for his community.[2] Endo and his friends spent many hours picking wild mushrooms and plants from the forests near the farms; this work was sustaining, and fateful.

In Japan, as elsewhere, some mushrooms can kill. Others are so prized, so full of delicate flavor, they command thousands of dollars per pound. Endo learned about the mushrooms in order to tell the delicious from the deadly; he learned about them in order to eat. The distinctions among kinds of mushrooms can be subtle — a difference in spore color, gill formation, stem shape and size, or simply preparation — but these distinctions are important, the difference, it turns out, between sustenance, medicine, and death.

Fungi cannot move. Ever since they colonized land hundreds of millions of years ago, fungi have been at the mercy of the conditions into which they are born and so have evolved an incredible arsenal of weapons of self-defense. They can only fight; flight is not an option. As a boy, Endo did not learn the names of these chemical weapons, but he learned their consequences. Years later, he would recall a day when his grandfather went with him to gather a mushroom called, in English, fly agaric (*Amanita muscaria*).[3] His grandfather released a handful of houseflies near a pile of these mushrooms. The flies flew toward the pile, attracted to its earthy smell. Once there, they began to eat. The flies gorged and then, one by one, died. His grandfather then took those same mushrooms and boiled and ate them. What seemed innocent could be deadly, and what was deadly could be made edible, delicious even. (Though it is worth noting that it is possible Endo's grandfather was making a mistake. Japanese mushroom guidebooks list fly agaric as poisonous even once cooked.) These local secrets spoke to broader mysteries that attracted Endo the way the fungus calls to the fly.

Even at the age of eleven, Endo knew that he wanted more than what could be harvested from the place he was growing up. He wanted to go to high school and then college; he wanted to do something grand. Endo wanted to heal people. The more Endo persevered, the more he faced new limits of his circumstances. His grandfather was the closest thing to a doctor in his town, and yet he was incapable of curing most diseases. Many in the town got sick and, simply, died. Of what, no one could say. When Endo was in fourth grade, his grandmother developed a cancer that none of his grandfather's local remedies could cure. Endo stood beside his grandmother's bed and felt the lump in her stomach, with its hard diameters. Before his grandmother died, he held his hands on her. Here was a moment when he needed magic. He had none—not yet, anyway.

Endo left his village at the age of seventeen to go to high school in the larger city of Akita. There, his interests in medicine and fungi intensified. It was more than professional interest; it was an obsession, one that led him to revisit the fly-killing mushroom. Endo wanted to know more about the poison in it. It seemed to be the boiling that took away the poison. Endo wondered if the boiling water simply washed away the poisons. It was an overly simple idea, but he was, after all, still in high school. Endo boiled some of the fly-killing mushrooms and then put the mushrooms out on one plate and the broth on another. Flies came to both, but only those flies that fed on the broth died. Clearly, the toxin was being washed away. These are the preoccupations of an unusual boy,[4] and yet it was out of the germ of precisely this sort of preoccupation that his greatness fermented.

When the time came for Endo to apply to college, his parents did not want him to go. They could not afford the tuition and sent Endo's older brother to talk him out of applying. Endo was unconvinced; he would pay his own tuition. He left the village for the Agricultural College of Tohoku University in Sendai, one of the seven Imperial Universities. Once there, he often went hungry. He would go to school late in the day to eat the food that others had left. Endo, as he would later say of himself, "always had his chop-sticks ready" for the bell that would announce that the poor students such as Endo could come eat what the wealthier children had left on their plates. Sometimes, even after these extras, Endo was still "too hungry to concentrate and could not hear a lecture," so close was he to fainting.

Upon his graduation in 1957, it would have been natural for the struggling Endo to try to work in a well-funded scientific field, one where jobs were likely, one in which struggle might be unnecessary. Had Endo listened to his parents and common economic sense, he might have chosen the well-trodden road, but Endo wanted to do something new, which, he seemed to know even from an early age, required him to choose the more difficult of each pair of options he would be given. The hard task he set for himself was to look to nature, to biodiversity's wild pharmacy of fungi in particular, for potentially useful compounds like those that had killed the flies. Few in the world worked in this field, but as a Japanese scientist, Endo had the advantage of his childhood experience with wild species, particu-larly mushrooms, an experience common in rural areas and also reflected in the scientific knowledge of Japanese mushrooms.

His first job was at the Sankyo Company in Tokyo. Tokyo was an enormous city, quickly becoming among the largest and most sophisticated in the world, but it had still not yet risen from the ashes of war. There, Endo was to study the enzymes of wild fungi to see if any might break down the pectin found in fruit. Pectin from grapes contaminated wine and ciders, making them bitter. If some-thing could break it down, that something would be of huge value.

Thinking like an evolutionary biologist, Endo had decided to search for such a fungus on wild grapes, where he reasoned that fungi would need to evolve compounds to break down grape pectin if they were to use all of the nutrients in grapes. Within a year, he found a fungus, *Pilidiella diplodiella*, that worked. Sankyo quickly commercial-ized the active compound from the fungus. The endeavor of learning from wild nature could bear fruit! Endo had tasted the flavor of his accomplishment. But he wanted more. He knew he would look to the fungi, but he wanted to move away from food. He wanted to work in medicine and save people—not just luxury drinks—and he thought he knew a way: by curing the problem associated with high levels of cholesterol in the blood by lowering those levels.

Cholesterol is a sterol, a kind of natural steroid, in the family of testosterone and cortisol, and it is found in cell membranes and, in animals, in blood, wherein it is moved from one place to another in the body. Nearly all organisms, from bacteria to dogs, need cholesterol. It is particularly vital to brain function; each of your thoughts requires a little. Yet, although cholesterol is necessary, beginning with work in Russia in the early 1900s, studies on rabbits had begun to suggest that too much cholesterol could lead to atherosclerosis. It would take decades and many more studies before these results were taken seriously, and yet, by the time Endo was doing his work, high cholesterol was viewed as a key feature of atherosclerosis, even if not yet a well-understood one. Perhaps if cholesterol could be lowered, atherosclerosis could be prevented. Ironically, blood cholesterol levels in Japan were, at the time Endo began his work, among the lowest in the world, but Endo imagined that as Japanese lifestyles became Westernized, the Japanese would be at risk too.

While Endo was still working on fruit and searching for new and useful fungi, a key step was taken in the study of cholesterol that would set the stage for Endo's idea to take full shape. Konrad Bloch at Harvard University worked out how the body makes cholesterol and saw that while some of the cholesterol in blood and cells comes from diet, most of it is produced by the body itself. In the liver, the enzyme HMG-CoA reductase carries out the slowest step in the thirty-step process of cholesterol production: altering HMG-CoA. This is the step every other enzyme must wait for, impatiently. Speed up that step, and more cholesterol is produced; slow it down, and less is produced. When someone eats more cholesterol, the system is naturally balanced, as the excess of cholesterol (in low-density lipoproteins, LDL) prevents HMG-CoA reductase from doing its job, and as a result, the body produces less cholesterol. Because of this balancing act, when lower levels of cholesterol are consumed, there is very little effect on the levels of cholesterol in the body; it is only when large amounts of cholesterol are consumed that the body's ability to regulate cholesterol levels in the blood becomes overwhelmed.

Individuals vary greatly in the levels of cholesterol their bodies try to maintain, and much of this variation is purely genetic. In those who suffer from the genetic disorder familial hypercholesterolemia, cholesterol levels tend to reach about 800 mg/dl if an individual has two copies (one from each parent) of the high-cholesterol gene variant; 300 to 400 mg/dl if he or she has just one. These high cholesterol levels are due to the relatively weak regulation of HMG-CoA reductase; the enzyme acts largely unchecked. Individuals with this kind of hypercholesterolemia are treated as diseased, but they really represent one extreme of the general phenomenon whereby individuals vary in their cholesterol levels (independent of diet) due to the versions of genes they've inherited from their parents. Endo did not understand the relative influences of diet and lifestyle on cholesterol yet, but he knew enough to realize that even once such influences were understood, the ability to chemically alter cholesterol levels in the body would be useful, lifesaving even.

Upon reading about Bloch's work, Endo immediately wrote a letter to him. He proposed to Bloch that he would discover a drug that lowered cholesterol, and he described the details of his plan. Since his childhood, Endo had imagined working in the United States, where his boyhood idol, the Japanese scientist Hideyo Noguchi, did his great work discovering and documenting the syphilis bacterium, *Treponema pallidum*. But Bloch—who had just won the Nobel Prize—did not answer Endo. He was a busy guy. One suspects that Endo was not the only promising young scientist who'd sent a letter. Instead, Endo went to Albert Einstein College of Medicine in New York City to work with Professor Bernard Horecker. There, Endo studied lipopolysaccharides in the cell walls of bacteria for two years before returning to Japan to work, once again, at the Sankyo Institute. The research in New York was not what Endo dreamed of; it taught him new skills he needed, but he felt as though he was biding his time, waiting to figure out how to lower cholesterol using fungi. If anything, the time in the United States, where so many people seemed to be dying of the consequences of high cholesterol, emboldened him. So he returned to Sankyo, eager to get started, but there were roadblocks. It took him three years to convince his supervisors that the idea he had for a big project on cholesterol, the same one he had originally written Bloch about those years before, was worthwhile.[5]

Most of the research that built on Bloch's discoveries focused on understanding the fate of cholesterol in the body or on the contribution of diet to cholesterol. But while diet and lifestyle were important, genes (and culture) seemed, at least to Endo, to be more important. If this was right, cholesterol levels might be lowered with reasonable diets and changes in lifestyle. But for those individuals who had a greater risk of high cholesterol and heart attacks in the first place, it required something more: a rebalancing of the equilibrium level of cholesterol. This is just what Endo wanted to do.

The need for more than just diet and exercise to deal with high cholesterol is widely discussed now, but it was relatively ignored for many years. In focusing on pharmaceutical approaches to controlling cholesterol, Endo again chose an unusual direction. He wanted to do something radical; just as with the pectin problem, he would take an evolutionary perspective. He made several intellectual jumps that no one else would make for years, and, at least in his retelling, he made these jumps more than five years before he could implement his ideas.

He decided to focus on slowing the production of cholesterol by the liver. To do so, he would search for some natural compound that would stop the key enzyme, HMG-CoA reductase. Endo's theory for how to do this was inspired by his childhood forays among the fungi and the work of Alexander Fleming, whom he read about in college. Alexander Fleming was both a scientist and an artist of sorts. He dabbled and he poked. One of the things at which Fleming dabbled was bacterial art. He would paint on bacterial culture plates with different strains of bacteria in such a way that they would grow at precisely the right rate (and produce the right color) so as to make, for a moment, an image—be it a baby, a policeman, or something else. In order to produce this art, Fleming needed a precise understanding of the growth rates, growth forms, and colors of different bacterial strains. This meant that Fleming was always on the lookout for unusual bacteria. One way to find them was to leave plates out to grow longer than he should. One time, in 1928, when he did just this, some microbes grew on his plates in a circle, and around that circle, there was a halo of, for lack of a better term, death. Within that halo, nothing seemed to be growing. The microbe in the middle was the fungus that produces penicillin, a species of *Penicillium*. The ring around it was the ring of antibiotics it produced to kill off the *Staphylococcus* bacteria with which it competed for food. *Penicillium* kill by releasing compounds (beta-lactams) that prevent bacteria from producing peptidoglycan, the mortar in their cellular walls.

Fleming had discovered penicillin, but also, in the process, the ancient wars between bacteria and fungi. Nearly all fungi that have been studied in detail now appear to produce at least some compounds that they use to kill bacteria (fungal antibiotics still account for 65 percent of those on the market). If there are one million fungal species (no one knows even roughly how many there might be, but this seems a reasonable guess), there are, buried in their bodies and genes, as many kinds of antibiotics, antifungals, and other useful drugs. We need antibiotics; we thrive thanks to them. But Endo had a broader insight, one that no one else seemed to have had—that one could use the antibiotics produced by fungi for ends entirely unrelated to killing bacteria. To get started, all one had to do was think about how fungi might attack bacteria and consider how that attack might be put to use.

Among the simplest ways for a fungus to destroy bacteria is to attack their cell walls (as with penicillin) and, inside those walls, their cell membranes. Cell walls hold bacteria together and keep them from diffusing into oblivion. Cell membranes serve as permeable filters. Endo knew that the cell membranes of many bacteria, unlike those of the fungi, were built on a scaffolding that included cholesterol. What if, he speculated on these plausible evolutionary grounds, some fungi had evolved ways to stop the production of the cholesterol bacteria required in their cell membranes?

Endo would search for such a fungus. The idea that it might exist was reasonable—at least with hindsight—but the possibility that he would find it was far-fetched. He was searching for a compound he had speculated might exist, but he didn't have any real guess as to which species it might be in. He decided to study a few thousand species of fungus. If he needed to, he would study more. He would spend years, if necessary, even decades. The only real limit was how long his employer would give him to search and, of course, how many kinds of fungi he could find to test. The task required many steps. Each fungus had to be found in nature and then grown in broth in giant flasks (which required knowing what to feed it). The flasks would then be stirred up with the liver enzymes (from rats) necessary to produce cholesterol.[6] Fungi able to stop the enzymes would then be studied some more. Among those, unsafe fungi would be discarded. Expensive fungi would be discarded. Fungi that were slow to reproduce would be discarded. Fungi that broke down when heated would be discarded. Then, when he had a list of fungi that passed all these tests, Endo would have to do even more tests to figure out what was going on, why the fungus seemed to work. Endo would double-check whether the rate-limiting enzyme animals use to produce cholesterol, HMG-CoA reductase, was inhibited. He would then try to isolate and purify the active compounds that were doing the inhibiting.

The work was tedious and grueling, less science than factory work, industrial microbiology. As Endo would say in one interview, it was "unlike a lottery, which must include prizes.... [In the search for drugs] no one knows if there is a prize." Each hopeful day was followed by a dozen hopeless ones. But finally, after screening thirty-eight hundred fungal species, a potentially useful antibiotic, citrinin, was found in the fungus *Pythium ultimum*. When injected into the bloodstreams of rats, citrinin inhibited HMG-CoA reductase and, more important, lowered cholesterol levels. There was a moment of ecstatic excitement in the lab, but then citrinin proved toxic to the kidneys of the rats. Back to the drawing board.

Then, one day in the summer of 1972, after another twenty-three hundred compounds had been tested, another compound was found that appeared to reduce the function of HMG-CoA reductase and, hence, cholesterol production. Once again, Endo and the lab were excited. Once again, they confronted a challenge. They couldn't isolate the compound in the fungal strain that was having the effect. Finally, in July of 1973, a breakthrough: they found the magical compound later named compactin. Compactin's mechanism of action was simple. It resembled the normal substrate to which HMG-CoA reductase binds, and this allowed the compactin to gum up the HMG-CoA reductase and prevent the production of cholesterol. In nature, this would allow the fungi to fight bacteria by preventing them from building cell walls. But what Endo hoped is that it might also allow him to fight heart disease by preventing cholesterol from becoming so abundant in the blood that it led to plaques.

The fungus from which Endo isolated compactin was found growing on rice in a grain shop in Kyoto; it also grows on fruits, including oranges and lemons. Compactin seemed, from the start, safer than citrinin. Maybe it was the compound he had been working toward. What was even more amazing to Endo was that the compound was produced by *Penicillium citrinum*, close kin to the fungus that had helped Fleming discover antibiotics in the first place. This alone seemed auspicious, but the key test was whether the compound could lower cholesterol levels in rats. Endo sent the compound to be tested at the Central Research Laboratories of Sankyo. He could do nothing but wait. Then the test came back. The cholesterol levels in the rats had not dropped. Endo's mind filled with despair and expletives. He tried again, this time performing the test himself on both rats and mice. Once again, no luck. As he would later say, "It looked as if two years of work and over six thousand tests had led nowhere."

At this point, many people would have given up. Giving up would have been forgivable; it would have been reasonable. Some discoveries are inevitable; the discovery of statins, the drugs that Endo would ultimately prove to be searching for and of which compactin was one form, was not necessarily one of them. But Endo did not give up; he picked himself up out of his exhaustion and tried to figure out what had gone wrong with the rats. Slowly, progress was made, and then, two years later—seven years after the idea for this project had originally occurred to him—he remembered the chickens! They were to be, he imagined, the chickens of hope. He had a friend, Noritoshi Kitano, with chickens. Perhaps, he thought, something about the rats he was using was unusual. The rats, he knew, had less cholesterol in their blood than humans did. Maybe the rats were weird. It was an extraordinarily remote possibility— after all, rats are the model for humans used in labs around the world, and the reason for using these model organisms is their similarity to humans. Humans and chickens are, well, different. But it was a possibility all the same, and the last unchecked Lotto ticket in a lottery that might not have a winner.

Kitano was about to kill his chickens, but over drinks at a bar Endo begged him not to. Endo knew how to take care of chickens. He soon scooped up his friend's chickens and spent the next weeks feeding hens and shoveling chicken shit while he tested them, one by one, and gave some of the birds his compound. On the thirtieth day, he went to check the chickens. He drew their blood. As he did, he wondered what on earth he was doing. If the compound did not work in rats, there was no reason it should work in chickens. But maybe, just maybe...Chickens do have high levels of cholesterol in their blood, much of which, at least in hens, ends up in their eggs. (Rats, by contrast, although Endo did not know it, have very, very little LDL cholesterol, the form that inhibits HMG-CoA reductase.) The chickens that had not been given the new compound had levels of cholesterol similar to those they started with. This was what had been expected. But what about those that had been given the compound? They would be the big test. What about them? In the experimental chickens, cholesterol levels had dropped by half! Half! Endo had just solved the cholesterol problem with a dozen chickens too old to eat. He would go on to repeat his experiment with monkeys, which were much more similar to humans than rats or chickens were. In the monkeys, the compound worked too.

The next steps (including testing compactin in dogs and in more monkeys) would prove challenging too, but in the end, Endo had achieved his dream of a chemical key to lock and close the body's cholesterol doors. What was left was going commercial. After all Endo had been through, commercial challenges might have seemed trivial. They were not. First, there was a residual worry: Was this drug safe? One study in rats (in which compactin didn't even work) suggested the drug made crystals form in the kidneys. It took Endo a year to show it did not, but Endo's company insisted that the compound did. They then argued (using data that have still never been shared) that compactin caused tumors in dogs. Endo's project was discontinued. Endo appealed the decision and tried every other route he could, but to no avail. It was over; all of his work was for naught, at least for Endo.

Endo's work on compactin encouraged others to search for statins, in some cases clandestinely. The drug company Merck had obtained a sample of mevastatin and unpublished data about it in 1976 as part of a disclosure agreement. Merck would not be so bold as to sell mevastatin, Sankyo's drug, Endo's drug, but it was sufficiently bold to study mevastatin and then look for similar drugs. Lovastatin (Mevacor) was isolated from the fungus *Monascus ruber* and then also another fungus, *Aspergillus terreus*. Lovastatin was approved by the FDA in the fall of 1987 and became the first statin drug marketed in the United States. Later, synthetic statins were produced by chemical analogy with the biologically produced versions or by simply altering the natural versions, synthetic statins that have led to Lipitor and Crestor, among other drugs, which are now collectively the most profitable type of drug on the market. As for Endo, he never benefited financially from his discovery. Endo left Sankyo, unceremoniously, in 1978. He took a job at the Tokyo University of Agriculture and Technology, where he has continued his research ever since with modest funding and no direct benefits from his statin research. As for his own cholesterol, he took Mevacor, Merck's drug, for a while but then stopped, choosing to exercise more, he says, instead. Meanwhile, the statins themselves have withstood their subsequent tests. Tens of billions of dollars of statins are now sold globally each year. More than thirty million people are now taking statins, and, as a result, tens of millions of additional years of lives have been lived.

That statins save lives, particularly in older people, whose arteries are more likely to have begun to clog, is unequivocal. Where they have come under debate is in terms of just how universally they should be used (whether they should be used in the young, for instance, or in those with other medical conditions). In addition, recent research has suggested that their benefits might be due to both reductions in cholesterol levels and an anti-inflammatory or even antioxidant effect. Yet, in the complex story of the heart, statins are about as close as one gets to an unambiguously beneficial treatment. They were made possible because Endo was willing to spend thirty years of his life thinking like a fungus, thinking evolutionarily. Between Fleming and Endo and a solitary genus of fungus, a single experiment has already contributed more to the survival of humans than all medical technology combined. How much more do the other three hundred species of *Penicillium* have to offer? No one has fully checked. No one has even yet named most of the species of *Penicillium*, much less studied them. In my lab alone, we have found dozens of new species of *Penicillium* fungus living on and in houses, and tens of thousands of unnamed species of fungi more generally. How much more magic do these and the other hundreds of thousands of unnamed and unstudied fungal species have? The amount is large but unknown. Since 1971, the study of wild animals, plants, and, especially, fungi and bacteria has yielded 70 percent of all discoveries of new, medically useful compounds. These include antibiotics, antifungals (fungi must fight other fungi too), and even cancer treatments, in addition to statin drugs. But most species on Earth, tens of millions of species, have not yet even been named, much less studied. Medically important discoveries wait in each scoop of soil; they grow on each piece of moldy bread and in every rotting leaf or log, anywhere ancient lineages wage war or coordinate peace with chemicals, anyplace where anything is alive.[7]

12

The Perfect Diet

kira Endo sought to lower the cholesterol levels in the
body through evolutionary chemistry. Though one can
debate how often the drugs Endo discovered should be
used, it is clear his insights yielded sweeping consequences. Many of us will live longer thanks to Endo's research. But
from the very beginning of cholesterol's story there has also been
another approach to lowering it, one that many view as more
natural—diet. The story of heart disease and diet is towered over
by one figure, a kind of complicated forefather who, depending on
your perspective, is either the hero or the villain of our modern
dinner tables: Ancel Keys.

Keys was born in 1904 in Colorado Springs, Colorado. When
he was a small boy his family moved to what they hoped would be
the prosperity of San Francisco to look for jobs; then the great
earthquake split the ground open. The family moved again, this
time around the bay to Berkeley, California. Even as a child, Keys
seemed marked for some form of greatness. A study of child
geniuses labeled him as one of the true young intellects. Perhaps
because of his intellect, he was restless. He wanted to explore the
world on his own terms, leaving for adventures even before he had
graduated high school. He worked in a lumber camp, shoveled bat
guano, and traveled to Asia on a ship on which he worked as a
mechanic. This wanderlust was punctuated by the moments of

stability during which Keys would achieve important milestones. Between trips, he started and finished an undergraduate degree. During a later period, beginning in 1927, he enrolled in the PhD program at the Scripps Institution of Oceanography. There, he began to study fish. Among their feathery gills and puckered mouths, the flower of his real genius began to unfold, a genius for big scientific projects, adventures in the unknown. It was this mature genius that would eventually turn toward the study of human bodies, diets, and hearts.

But not at first; he still had to figure out the fish. At Scripps, Keys focused on physiology and how fish—especially killifish—dealt with hypoxia, the absence of oxygen, an absence felt first in the active muscle of their hearts. It was in studying hypoxia that he became interested in the body's powerful ability to regulate itself, its tendency toward homeostasis. After finishing his PhD, during a brief stay in Copenhagen with the Nobel laureate August Krogh, he built on his work with killifish to understand how eels could keep the salt concentration in their blood the same as they moved from salt water to freshwater and back. Keys readily made new discoveries, the ease and frequency of which inspired in him a complicated mix of humility (at the grandeur of what we don't know) and boldness. Had he stopped all research right there, he would still be remembered, but he did not stop.

In 1933, Keys returned to the United States and decided to try his hand at studying humans as part of the Harvard Fatigue Lab. How different, after all, could humans be from fish? Keys was to be part of a growing number of scientists who had become interested in the ability of the human body to maintain its normal function under difficult conditions (much as fish do in different bodies of water). Keys was particularly interested in studying the body's response to high elevations. Toward this end, he led an expedition to observe the acclimation of bodies (including his own) to high elevations, an expedition that allowed him to simultaneously satisfy his wanderlust and advance science. The expedition, in which Keys led a large group of scientists, was a collective self-experiment, something very common for exercise physiologists, particularly those at Harvard. The scientists would measure how their own bodies responded to high elevations. Predictably, the journey was not without challenges (including a high-elevation fistfight), but it was a scientific success.[1] On his return, Keys wrote articles about the size of athletes' hearts (much larger than normal), the properties of insulin, the chemistry of blood, the exchange of carbon dioxide between mother and fetus in goats, the permeability of capillaries, the ways in which oxygen moves from blood into tissues, the effects of testosterone and estrogen on the body, and much more. The papers were insightful, bold, creative, and, like him, wandering. For this work, he was lauded and rewarded. The largest of the rewards was (after his time at the Mayo Clinic, where he met his wife, Margaret), the establishment, in 1937, of his own lab, the cryptically named Laboratory of Physiological Hygiene at the University of Minnesota.

All of the studies Keys did, at his own lab and elsewhere, would help him in his ultimate quarry, the study of human hearts and cardiovascular disease. The high-elevation study taught him how to lead lab scientists into the wilds of the field, for example. But perhaps the most useful (and foreshadowing) experience Keys had was the challenge of developing a perfect military food. The military was unhappy with the food that paratroopers were getting. Keys ran a physiology lab. He thought himself sufficiently clever and experienced to do the job, and so he went to the Quartermaster Food and Container Institute for the Armed Forces, in Chicago, and suggested to those in charge that he could solve their problems. They declined his help, but he was able to get support from the Cracker Jack Company in the form of Cracker Jack boxes and a small amount of money from William Wrigley's office (the same Wrigley of spearmint-gum fame). With the Wrigley funds, Keys bought things from the local supermarket and tried to produce a meal that, in light of his other studies on vitamins and energy requirements, was both balanced enough to be sustaining (given the knowledge at the time) and tasty enough to prevent revolt. He wanted to make the food healthy. He was not yet thinking about the health of the heart (though that would come), but he already knew that success with diets needed to be equal parts taste, marketing, and science. The approach worked. Keys produced what would later be called the K ration, a package of food items intended to be used only by paratroopers in emergencies. But military leaders liked it so much (General Patton thought it a military breakthrough) that it became the standard ration given to all fighting men. It was so successful it grew to be normal.

Keys seemed to be able to do anything he set his mind to. Not only could he make discoveries; he could also convince others to follow his lead. Just how far others would follow him was about to be tested. Keys decided he wanted to figure out how to "predict both short time and long range physiological results from a given mode of life, personal habits, activity, and diet."[2] This would require experiments with humans, experiments more extreme than those he was willing to conduct on himself.

The most obvious immediate need in terms of understanding the influence of diet was in the context of war. World War II had reached its most awful moment. Soldiers of many nations were starving, as were civilians, but little was known about what happened to the body during such conditions or how to heal the body of someone who had been starved. Keys decided to enlist conscientious objectors, many of them Quakers, to be part of a study on starvation. The men gathered in the basement of the football stadium at the University of Minnesota where Keys had set up his lab. (The space was ample, though it vibrated with roars each time a touchdown was scored.) There, they were subjected to a variety of low-calorie or starvation diets. Men grew so hungry they became desperate, but the shame of leaving the project was akin to the shame of abandoning a loved country, and so no one left. Personally, Keys was torn about the trials he was inflicting upon these men. Yet, overall, the men themselves believed in their mission, even when it became most difficult. In later years and with very few exceptions, the men in the starvation trial said that if they had it to do over, they would participate in the study again. The privations they experienced were extreme (one man chopped his fingers off with an ax in order to go to the hospital and get out of the study without seeming weak), but so were the benefits their volunteerism offered society.[3] The study resulted in the single most comprehensive treatise on starvation ever published. Medics used the results when starving concentration-camp prisoners were freed and needed to be fed. It is used today in treating individuals with extreme eating disorders. Keys wanted to do work that lasted and that mattered, and he succeeded.

From his work on fish, K rations, and starvation, Keys came to contemplate something even more central to postwar humanity: he wanted to understand what everyone should eat, not just soldiers or starving people. To Keys, diet and lifestyle were a unifying feature of his research, even when he was studying eels or killifish. And, at least for humans, cholesterol seemed to be a key player in the story of diet, lifestyle, and health. Cholesterol, he knew, was associated in some way with atherosclerosis and heart disease. Keys wrote about the cholesterol in cow's milk and its potentially negative effects as early as 1948. Initially, Keys did not know that there were different types of cholesterol.[4] Nor did he know that the cholesterol humans consume has a relatively modest effect on the cholesterol levels in the blood, but he was clever enough to make his way.

Everywhere Keys looked, he saw the importance of better understanding cholesterol, atherosclerosis, and heart disease. Around Keys in Minnesota, in the years after the war, heart disease seemed to be epidemic among business executives, and Keys well knew that those executives consumed, as if by law, rich diets dense and slippery with cholesterol—eggs, butter, milk, more butter. Keys, among others, thought the extent of this phenomenon new and troubling. Keys stalked the obituaries and found them full of wealthy, powerful men having heart attacks.[5] In 1900, pneumonia was the leading cause of death, and by the time Keys was checking the obituaries, in the early 1950s, heart disease seemed to have taken its place.

Keys came to believe that consuming more animal fat caused the body to produce too much cholesterol, which, in combination with the consumption of cholesterol itself, led to higher rates of atherosclerosis. Such a link had not yet been documented in the lab or among humans yet, but with his hypothesis in hand, Keys envisioned an informal test. He noted the Minneapolis businessmen he saw dying all around him and decided to observationally study 286 of them. He would give each man a physical examination, including blood tests for cholesterol levels, a few times a year for what would ultimately be twenty-five years. He also tracked their deaths. Keys could be persuasive. The men agreed. He had convinced conscientious objectors to starve themselves nearly to death in the name of the war against Germany. Now he just had to convince men to submit to blood tests and the like in the name of the war on heart disease. The men who agreed, including Bernie Bierman, the famous University of Minnesota football coach, and Edward John Thye, then the governor, were all excited to be part of Keys's vision.

Keys expected to see higher blood cholesterol levels in those men with diets richer in animal fat and cholesterol; that is just what he found. The men with diets richer in animal fat had more cholesterol in their blood. Because more fat was associated with more cholesterol and more plaques, Keys began to wonder if a change in diet could reduce cholesterol and plaque formation and therefore potentially prevent heart disease.

Keys's data suggested that a simple dietary change such as

reducing saturated fat intake might decrease heart-disease risk, but he needed more direct evidence. For one thing, while there did seem to be a link between fat in the diet and cholesterol in the blood, maybe there were other factors at play. Also, the cholesterol levels in the businessmen, who had high rates of heart attacks, were not any higher than those in the general public. Keys decided to do an experiment in order to really isolate the effects of fats. He gained permission to manipulate the diet of thirty patients at the Hastings Hospital for the Insane. Fifteen patients were given diets rich in animal fats, akin to what patients who weren't in the experiment received; fifteen were given diets in which the animal fats had been replaced with starches. The cholesterol in the individuals eating the animal-fat diets went up, while that in the individuals consuming the starches went down. Keys was as convinced of his idea as ever, even though in retrospect, the change appears very modest— statistically significant but perhaps not practically so. Cholesterol in the blood increases when animal fats are consumed. The cholesterol shepherds the animal fat, which isn't soluble in water, through the blood. That the levels of cholesterol in the body increase when one eats animal fat is not, in and of itself, a malfunction. It is the body's marvelous response to a hard-to-move compound, but somehow, Keys felt sure blood cholesterol levels must be associated with the extent to which high cholesterol leads to atherosclerosis.

The executive study and the small experiment were useful in informing Keys's intuition. They convinced him he was on the right track, but he needed something grander in order to convince the public and his colleagues. The ideal would have been to do a global experiment in which he assigned people around the world to one of several randomized diets and lifestyles. Such an experiment was not and will never be possible, not even for Keys.[6] But Keys was going to try the next best thing, a pair of "natural experiments."

Keys's first approach was to study what happened to people from countries around the world when they move to a new region with a new culture; Hawaii, in this case. This is the sort of natural experiment that evolutionary biologists and ecologists do. The British ecologist Charles Elton pioneered one version of this approach. He realized that when species from one place were put in another, they represented, however unintentionally, a test of ecological theory. Scientists could observe which species succeeded where and how fast they evolved. Keys, thinking again like an ecologist or a physiologist, saw immigrant humans in the same vein. These individuals from around the world would all be subjected to the diets of their new places, but they came from very different genetic, cultural, and dietary backgrounds. In theory, Keys could observe not only the influence of Hawaiian-American life but also whether that influence affected different peoples differently.

Japanese immigrants to Hawaii were an interesting test of this hypothesis because Hawaiians, like other Americans, consume much more animal fat than do people in many other regions of the world, including Japan. Sure enough, when Japanese people, with an average cholesterol level of 120, moved to Hawaii, their cholesterol levels rose to about 183. Their risk of heart disease rose too. Similarly, when Japanese people moved to Los Angeles, where their diets were even more Americanized, their cholesterol levels increased to 220, and their rate of heart disease increased as well, even relative to that of Japanese people in Hawaii. This evidence was compelling, though it had several problems, the biggest of which was that becoming Hawaiian or Los Angeleno meant many other changes in addition to the shift to a diet with more animal fat. Diet is enmeshed in culture, and moving to a place means becoming part of it. A fattier diet was part of that culture, but so were many other things, such as more hours spent indoors, more smoking, a lack of vitamin D, increases in salt and sugar intake, altered exposure to parasites, and increased exposure to pollution, to name a few.

Beginning in the late 1950s, Keys decided to do another natural experiment. He would compare the diets of people living in many different countries and, in turn, their rates of atherosclerosis and heart disease. His first two countries were Japan, where the diet is based on fish and rice with little animal fat, and Finland, where, as Keys would put it, people sometimes dined on meals consisting entirely of butter and cheese. When Keys compared these two regions (which he did by training doctors in both countries to perform a standardized set of repeated measurements on patients; once again, Keys was convincing people to work for a common goal, and once again, it was his goal), he found that in the country where people ate more fat, Finland, there was more heart disease. Keys then met more people and enlisted more countries in his study. By 1956, Greece, Yugoslavia (sites now in Croatia), Italy, and the Netherlands had signed on, leading the project to be called the Seven Countries Study; there were seven countries but even more individual sites.

The Seven Countries Study required Keys to fly around the world again and again coordinating activities. It was not easy. Each data point was hard fought, and there were so many data points; Keys and the vast team he brought together worked with more than ten thousand people (as he would later say, from Bantu tribesmen to Italian *contadini*). The results, and even the endeavor itself, were controversial, but Keys was, as ever, compelling. He began to gather endorsements—for lack of a better word—from influential people, such as Paul Dudley White, President Eisenhower's personal physician. With White's important blessing, the study gained more traction among doctors and scientists than it might have otherwise; it also gained funding. The study still continues to this day (minus a few countries that weren't able to find funding to continue), but for Keys, the big result came after the first five years, at a point when he plotted a clean line on a graph showing that the countries where people had the most animal fat in their diets were also the countries where people had the most deaths due to heart disease, heart attacks in particular. Finland had the highest incidence of heart-attack deaths, 992 out of 10,000 deaths, and its people consumed the most saturated fat, animal fat. People in Crete, with their lifestyle of sun, relaxed days, and diets rich in vegetables and olive oil, had lower blood cholesterol, and just 9 deaths out of 10,000 were due to heart attacks. These patterns have held up over the years. As of twenty-five years after the beginning of the study, the number of deaths due to heart disease in a region was still strongly positively associated with saturated-fat consumption (particularly with butter consumption). Also, time added another insight—those deaths tended to be negatively associated with the consumption of olive oil, fish, and wine. The take-home, in its simplest form, was this: eat less saturated fat, eat more olive oil and fish, drink some wine.

Based on their results, Ancel Keys and his wife, Margaret, who had by this time become intimately involved in both Ancel's research and the marketing thereof, could simply have written a scientific paper. That would have been the standard next step, but Ancel Keys was not a standard man, nor were he and his wife a typical team. They wanted something bigger. They had, they thought, figured out a way to prevent atherosclerosis by lowering cholesterol through a change in diet. The original data from the Seven Countries Study were not data on atherosclerosis or even cholesterol— they were data on heart disease and fat intake—but for Keys, the link between the two was sufficiently compelling for him to advocate change. Most scientists never advocate anything; they never make suggestions for fear of not knowing quite enough, not yet. Keys would not be one of those scientists—he would hold forth, he would recommend.

What the Keyses came to advocate was, in essence, a K ration for America—a diet that would get Americans through not war but their daily lives. The Keyses focused on the Mediterranean diet—

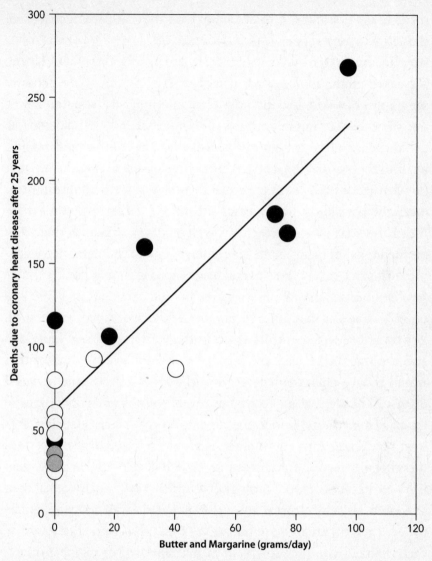

The relationship between the number of deaths due to coronary heart disease in the Seven Countries Study and the average quantity of butter and margarine (two key contributors to saturated fat) consumed in each region. After accounting for the effects of saturated fats, countries in which individuals consume more olive oil (white circles) or fish (gray circles; Japan) have fewer deaths due to coronary heart disease than would be expected based on their saturated fat alone.

something along the lines of what Southern Italians consumed, though with any vegetable oil taking the place of olive oil and other simplifications to make things easy. It was not the only diet that the Keyses thought might work (the Japanese in the study actually seemed to have the lowest cholesterol and longest lives), but it was one they thought might both lower cholesterol and spread person to person, engendering financial success for the Keyses and public-health success for humanity. While the Keyses' version of the diet had features of the Mediterranean diet, it was not a Mediterranean diet, and it certainly did not entail an entire Mediterranean lifestyle. The Keyses had to assume that the elements they kept in their rec-ommended diet maintained the region's central character: its neces-sary influence on cholesterol and, they thought, heart disease. Their first book was *Eat Well and Stay Well*, which came out in 1959. The book, unlike any diet book ever written before or since, began with a two-hundred-page introduction to the work the Keyses had done and the science of diet and heart disease, followed by meals that would reduce cholesterol levels and the risk of death due to heart disease. The work was so compelling and so fully embraced that the lessons from the book became a part of how Americans ate, or at least how Americans tried to eat. It also became a part of how doc-tors thought about heart disease. In 1959, before *Eat Well*, the Amer-ican Heart Association took the position that the link between saturated fat, cholesterol, and atherosclerosis was not well estab-lished. In 1961, after the publication of *Eat Well*, and with Keys by then on its board, the American Heart Association advised Ameri-cans to avoid saturated fat and cholesterol. Eschew, they and Keys said, butter, meat, egg yolks, and full-fat milk. The American Heart Association began to recommend a diet that bore a striking resem-blance to that which the Keyses advocated in *Eat Well*. That same year, Keys was on the cover of *Time* magazine, portrayed as the genius behind the modern, scientifically informed diet.[7]

The Keyses' book ushered in a change in the American relationship to food; if you are alive today in America, your perspective on food has been shaped by the Keyses. The Keyses convinced Americans—including scientists and doctors—that saturated fat was bad. Animal fat was regarded as particularly bad, but all fat came to be suspect; this was a novel idea. Before the Keyses, people ate the food of their respective cultures. There was, of course, good food and bad food, but goods and bads were handed down from one generation to the next. There were no experts at the dinner table and few pronouncements heard on the news about what would kill you or allow you to live forever. The Keyses helped to create the seed of a culture in which the experts on what we should eat were scientists.

In retrospect, we can identify complexities associated with the story of diet and heart disease that the Keyses offered America, complexities people missed in their attempts to simply consume the advice as though gobbling up K rations. The first had to do with the science. Early on, Keys had become relatively confident about the relationship between dietary levels of saturated fat and cholesterol, blood cholesterol levels, atherosclerosis, and heart disease. The Seven Countries Study data supported what Ancel Keys had already come to believe through his observations and small experiments and those of others. The data were in line with what Keys expected and what many analyses confirmed, but, as later analyses have shown, they were just part of the puzzle, a puzzle with many pieces. The pieces the Keyses figured out seem to hold together, and they hold up to the scrutiny of time, but they are just the edge pieces of a very big puzzle.

In one unfinished section of the puzzle were more pieces of the cholesterol story. When the Keyses started working, there was just one kind of cholesterol. There is still just one, but that one type of cholesterol travels around the body in multiple forms, forms that were first identified by John Gofman, a physicist who worked with

Glenn Seaborg on the plutonium nuclear bomb and then went on to get his medical degree. What Gofman figured out was that the two most common ways of travel for cholesterol are LDL and HDL, low-density lipoproteins and high-density lipoproteins. LDL and HDL are familiar acronyms, but most of us think about them in the wrong way. These are not types of cholesterol; they are more like types of boats into which cholesterol is incorporated for the ride. Cholesterol and triglycerides are not soluble in water (or blood) and they can't go it alone through your blood vessels very well. High-density lipoproteins have a relatively big hunk of protein and relatively little in the way of cholesterol and triglycerides, and so they are dense, heavy for their size. Low-density lipoproteins are the opposite, with lots of cholesterol and triglycerides relative to the amount of heavy protein; they are less dense, light for their size. The body uses these two boats in different ways. Cholesterol in HDL tends to be removed from the blood and ferried to the liver, where it is broken down and excreted in bile. Cholesterol in LDL, by contrast, tends to be carried to the organs. Having more HDL boats can actually help reduce your LDL because they gather up the cholesterol left over in organs or in artery walls and drag it to the liver to be degraded. But it is more complex. Cholesterol appears to travel around the body in at least three other forms, and each of these forms plays a specific and only partially understood role. What is more, the real problem is not the LDL cholesterol itself. The real problem is that the LDL cholesterol, in bouncing around the body, gets damaged—oxidized—in much the way that a firm, fresh pear turns brown. The oxidization of the LDL leads white blood cells to swoop in, and it is the swooping white blood cells that initiate plaque formation. The white blood cells produce a kind of foam that is itself oxidized, attracting even more white blood cells. All of this is balanced, to a lesser or greater degree, by HDL, which can gather up the cholesterol in the LDL and drag it to the liver. When it does, the plaques can break up, the immune system having finally realized that everything is cool.

Your dietary and medical choices alter this complex system, both the pieces of the system we know about and those we do not yet understand. Reduce consumption of saturated fats, and you can reduce levels of cholesterol in LDL in the blood. Increase the amount of cholesterol you have embedded in HDL, and you reduce your levels of cholesterol in LDL in the blood. Reduce your inflammatory response, and you can reduce how often your body attacks oxidized LDL. Or up your antioxidants, and you can reduce the chances that your LDL will get damaged in the first place. There are other processes involved too, other areas of the puzzle. Although it was not understood during the time the Keyses were working, when individuals eat too much simple sugar (glucose or fructose, particularly in the form of high-fructose corn syrup), that sugar is converted in the liver into triglycerides, which hitch a ride with LDL and HDL, so sugar consumption can affect cholesterol levels and dynamics. Also, the chance that a plaque, composed of dead immune cells, triglycerides, and LDL cholesterol, in an artery will rupture is influenced by blood pressure (which is influenced by stress, alcohol consumption, salt, smoking, and more), and the chance that ruptured plaques will lead to clogged coronary arteries is influenced by how likely your blood is to clog. Then there undoubtedly exists complexities we have not yet figured out or are only just learning of. As a result, just because a particular people's diet is associated with heart health does not necessarily mean that one can simply switch to that diet and reap the full benefits of it; sugar consumption, inflammation, and genes also contribute to the health of the heart. Many bodily pieces are in play. There is more to being Italian than eating pasta and bread with olive oil.

Another problem was that the Keyses' messages became over-simplified in the media and even in the medical community, so the lessons the public took from the Keyses' inspiration were some-times dangerously inaccurate. The most pernicious example of this oversimplification is the case of saturated fat. Ancel Keys railed against the intake of animal fats because of the links with blood cholesterol levels, atherosclerosis, and heart attacks. Animal fats came to be described as saturated fats, to differentiate them from the unsaturated fats in vegetable oils (which, throughout the Key-ses' work, showed the opposite effect on cholesterol, atherosclero-sis, and heart attacks, a beneficial effect). To cater to the demand for unsaturated fats, the food industry began to create new sorts of fats that were both unsaturated and cheap. Fat molecules have long tails of carbon and hydrogen atoms. Different fat types differ in how the pieces of these tails are arranged. Fats with fewer hydro-gen atoms are said to be unsaturated and are usually liquid at room temperature (for example, olive oil). Saturated fats have more hydrogen atoms and tend to be solid at room temperature (think of lard). The newer fats were trans fats, which are unsaturated fats (liquids) to which some hydrogen atoms have been added (making them partially hydrogenated and changing the structure of some of the molecules). The Center for Science in the Public Interest (CSPI) and the National Heart Savers Association (NHSA), informed in large part by the work of Keys, advocated for trans fats as replace-ments for saturated animal fats. Trans fats, after all, were not ani-mal fats. The trans fats the CSPI recommended were primarily those in partially hydrogenated vegetable oil (typically soybean oil). Partial hydrogenation makes oils more solid and resistant to rot. It also, though the CSPI did not know it at the time, makes them more likely to contribute to the clogging of arteries (they simultaneously raise bad LDL and lower good HDL). For this rea-son, trans fats, which were created as a salvation, have recently been banned in more than a dozen U.S. jurisdictions and have been replaced in many thousands of food products. As I write, the U.S.

Food and Drug Administration is taking steps toward a wholesale ban of these fats.

The Keyses can't be held accountable for all the ways in which their work has been extended or misconstrued. But in one regard, they may have played a role in this process. The Keyses, Ancel Keys in particular, had an intuitive sense of marketing and of what people wanted. These were the skills that made the K ration successful. Ancel Keys was also capable of convincing people, including himself, of the greater good that could be achieved. This is what made men starve themselves for their country (but also for Ancel); it is also, in part, what convinced millions of people that they could, through conscious choice and scientific knowledge and reading the Keyses' book, better their health and lives. To reach so many people, Keys knew that the food needed to be tasty and the message needed to be, if not quite simple, then at least simplified.

But the big problem with what Ancel and Margaret Keys suggested was the great divide between what individuals strive for and what they achieve. The Americans whom the Keyses implored to change their ways were not conscientious objectors or soldiers; they could not be forced to do what they were told, even if they were convinced it was good for them. The average American is far less able to control his or her diet than the conscientious objectors in Ancel Keys's starvation study were. As a result, on average, Americans now weigh 50 percent more and have higher cholesterol (when untreated) than they did when the Keyses published their book. This is not because the diet the Keyses proposed made them fat and unhealthy. To the extent to which it has anything to do with the Keyses, it is because Americans followed up on those aspects of the Keyses' diet that were easiest. They, as Keys instructed, avoided fat and tried to reduce calories. But to do so, they switched to carbohydrates and the cheapest calories, processed sugars (which themselves contribute to heart disease). In switching away from

animal fats, they switched to the cheapest fats, partially hydrogenated oils (which are worse in terms of heart disease). Americans
also, despite Keys's instruction, ate more, and when one consumes
more calories (particularly calories in the form of simple sugars)
than one burns, one becomes, well, fatter. With this weight gain, a
person is maybe more likely to have arteries filled with bad cholesterol but definitely more likely to suffer from one or more of the
other maladies associated with obesity (or the awful combination of
all of those disorders: metabolic syndrome).

The trajectory toward eating too much of all the wrong things
has been a long time coming. One can compare the size of meals
from long before the Keyses to the size of meals in the present. In
the earliest paintings depicting the Last Supper, for example, the
size of the meal was half what it was a couple of hundred years later,
in Leonardo da Vinci's version. The table in his *Last Supper* is, in
turn, set with a meal considerably lower in calories than that in
almost any modern painting involving food (and notably devoid
of any meat except fish). The size of meals has risen in direct
proportion to the availability of food. One expects that the composition has shifted too, in accordance with the cheapness and tasti-
ness of foods. Two-thirds of Americans (and similar proportions of
Brits, Australians, and so on) are now overweight or obese, with the
proportion who are obese having doubled since the time of the
Keyses' book.

In other words, since learning ways to improve their diets from
Keys, Americans have gotten less healthy, specifically in terms of
heart disease, but also more generally. Many blame Keys for advo-
cating the wrong diet in the first place. In today's diet-book aisle,
you can find books that argue that to lower your cholesterol or live
more healthily, you should eat, variously, less red meat, more red
meat, less sugar, fewer carbohydrates, raw foods, more completely
cooked foods, food like our ancestors ate, or nothing but fruit.[8] Yet
all of these admonishments miss the point. Thousands of scientists

since Ancel Keys have spent their lives studying how people can eat better. Many times more nutritionists have implored us to eat one or another of those well-researched diets. Yet we have, from one day to the next, progressively consumed foods that are worse and worse: foods with more calories; foods with fewer key nutrients; foods more likely to cause us to die, if not from heart disease, then from diabetes; foods no one argues are good.

A recent Spanish study makes this point clear.[9] The study set out to examine the effects of versions of two Mediterranean diets — basically, simplified traditional Spanish diets — on individuals who usually consumed high-calorie, carbohydrate-rich, nutrient-depleted diets with large doses of red meat. Spaniards' diets have become so influenced by the global diet that the researchers could perform their study in Spain. In essence, they compared two diets like the kind Spaniards used to eat with the diet many Spaniards now eat. The scientists were able to convince participants to change their diets while going about the rest of their lives normally. They had one group of people, the control group, eat an average Western diet, essentially what they had been eating. They had another group of people eat a diet supplemented with olive oil, four table-spoons a day. Finally, they convinced a third group to consume extra olive oil, fatty fish, and tree nuts. The participants in this project were able to maintain these diets for two years. One can speculate that this is in part because the control diet was similar to one many people were already eating, and the experimental diets were tastier than the control diet. It must have also helped that high-quality fish, nuts, and olive oil were provided free by the study.

The study considered about seventy-five hundred people and focused on the effects of the different diets on heart disease and stroke. Overall, the study reported roughly 150 heart attacks, a similar number of strokes, and a similar number of deaths due to other heart-related problems. Amazingly, given the relatively short duration of the study and the modestness of the dietary intervention, it was found that individuals who consumed the control diet were at an increased risk of heart attacks, strokes, and heart-disease-related deaths relative to those consuming the olive oil–enriched diet and the nut and olive oil–enriched diet. Those consuming both nuts and olive oil were best off, particularly when it came to the risk of stroke.

One lesson that can be derived from this study is that versions of the Mediterranean diet really do seem pretty good for heart health, at least for Spaniards, with their genes, lifestyles, and microbes. Perhaps the same would hold if the experiment was repeated in the United States (though it might not, because of the many ways we differ from Spaniards in genes, lifestyles, and microbes). But the other lesson is that even people in Spain and Italy don't eat Mediterranean diets anymore.

As for the fate of the Keyses, on some level, it seems, they knew things were more complex than they had described. As they began to age, they faced the same choices any of their readers might have in terms of deciding how to live. They could have adopted the diet they proposed, or tried to anyway, and lived the Mediterranean way in Minnesota. Instead, they decided to move to one of the places in their study that had the longest life expectancy (and winters far more forgiving than in Minnesota)—Southern Italy. There they lived into their nineties (before ultimately moving back to the United States for the last few years of their lives), partaking of a real Mediterranean diet and reaping the personal benefits of a career studying the best conditions for a long, balanced life.[10]

13

The Beetle and the Cigarette

iet, statins, angioplasty, and heart transplants are largely separate, unwoven threads in the story of the heart. But if you enter the hospital with chest pain and someone uses an angiogram invented by Forssmann and Sones to see your coronary arteries and discovers that one of those arteries is narrowed, hardened, and partially clogged, what is next? Which innovation or regimen is for you and will extend your life or even just improve its quality? For most of human history, what you did about chest pain was: nothing. You waited for the Russian roulette of your heart to play out. No one could have figured out that you had narrowed or blocked arteries, and even if someone had, he would have been able to do little more than lament your sorry fate.

Today, if you go to the hospital in the United States with chest pain and are found to have a very narrowed coronary artery, you will be given statins, told to "eat well" in the future, and then, depending on the severity of your heart disease, very likely be offered angioplasty and a stent. In the United States, more than half of patients who go to the hospital with the strangling sensation of angina and are diagnosed with coronary artery disease are given stents (typically simultaneously with drugs and dietary and lifestyle admonishments). More than a million stent surgeries were performed last year in the United States, one new stent for every two hundred and forty adults. It is both the most frequent and the most sophisticated solution. Along with the stent, a remarkable variety of additional procedures can be implemented inside your blood vessels. Your doctor can start with a stent and improvise based on what he or she finds. Work inside blood vessels has become some miraculous cross between *Twenty Thousand Leagues Under the Sea* and what the Roto-Rooter guy does to your clogged drainpipe.

Elsewhere in the world, the options are slightly different, or at least chosen in different proportions. In Canada, for instance, angioplasty and stents are far less common than in the United States. One explanation might be that perhaps stents really are the very best answer most of the time but Canadian medicine chooses cheaper and less effective alternatives. Yet we know that life expectancies in Canada and the United States are similar. Another possibility is that for some reason, in the United States we are choosing the very expensive solution where the cheap one might do.

The proof should be in the pudding—the fate of patients—but before we get to the pudding, let's look at two scenarios. In one scenario, a patient comes into the hospital having actually suffered a heart attack. In this case, treatments are relatively similar in different regions and almost inevitably involve either a bypass or a stent. Far more common, though, is the scenario in which a patient comes in with angina and is found to have stable but severe atherosclerosis of the coronary arteries. In the latter case, a coronary artery might even be fully blocked, but if it is, blood is still finding its way through smaller, collateral coronary vessels, enough blood to keep the heart muscle working. A new series of studies provides as clear an answer as one could hope for, given the stumbling realities of science, medicine, and death, on what to do in this latter, stable situation. Before we turn to that study, let's make clear the main options available to a doctor when a patient comes in with angina and a nearly clogged (or truly clogged) coronary artery. He or she could do a bypass. These are still done by the thousands but tend to be performed only in the most serious cases—when a patient is actually experiencing a heart attack—or for those patients who are at high risk for any of a variety of reasons (patients with diabetes and more than one clogged coronary artery are particularly likely to benefit from bypass surgery, as opposed to medical treatment or stents). Bypass surgeries require a heart surgeon, and heart surgeons are becoming less and less common at hospitals. The descendants of the great heart surgeons of the 1950s and 1960s now face tough competition from the cardiologists.[1]

The doctor could perform angioplasty and then place a stent. There are different types of stents, though doctors do not tend to choose the "best" stent. They choose their favorite. As a patient, you do not typically choose which stent; you just choose *stent*, and the doctor goes to get the one she or he always uses.

Finally, the doctor could simply employ a noninvasive medical treatment that relies on statins, beta blockers, aspirin, nitrates (such as nitroglycerin), calcium channel blockers, a prescription for exercise and dietary changes, and fate. Other options also exist; there are heart transplants, for example, but these are done rarely—only a few thousand a year—and considering their cost, they and other more invasive heart surgeries are reserved for when things get really bad, when the heart is failing or has failed. We'll focus here on the most common alternatives.

Given these options, it seems important to understand how the outcomes for individuals with atherosclerosis and a bit of angina, that canary in the heart's most dangerous mine shafts, compare among potential treatments. Again, American doctors tend to consider stents the best option. But the factors that go into the choice of a medical treatment by doctors are complex and influenced heavily, if not always consciously, by incentives, available technology, and the concepts of disease doctors learn in medical school and that are inculcated into the culture of medical practice. We hope doctors spend extensive time reading scientific papers, learning the newest information, and following up on the fates of their patients. But many don't have the chance; their jobs are stressful and busy, and they have less and less time for pondering, while the literature on potential treatments gets larger and larger. Historically, doctors were trained to be doers[2] who learned through one anecdote or case study after another rather than via scientific evaluation of long-term results. A new generation of doctors is being pushed to weigh evidence and science. But the easiest place to learn about new science is still, as often as not, at big meetings, where those with the loudest, most sophisticated voices are not scientists but representatives of companies that design devices such as stents.

Many studies have compared stents and coronary bypass and found them to be similarly effective in situations where a coronary artery is blocked or nearly so (that's not the case if a patient actually experiences a heart attack and his condition becomes unstable or if he suffers from three or more clogged coronary arteries[3] or from diabetes). Functionally, this means that given the choice between those two options, doctors tend to choose stents, since the procedure can be done at nearly any hospital, whereas bypasses require heart surgeons (who are ever more rare, thanks to the rise of stents). Also, as one cardiologist bluntly put it, "Stents don't require you to have your sternum sawn in half." As a result, stents have become progressively more common while bypass surgeries have become progressively less common, the same general trend that has occurred with cardiologists relative to heart surgeons.

But just comparing stents and bypass assumes that one or the other of these two procedures is the best solution. A 2012 review in the *Archives of Internal Medicine* compared patient outcomes for stents (which, as noted above, tend to be similar in effectiveness to bypasses) with patient outcomes for noninvasive medical treatment

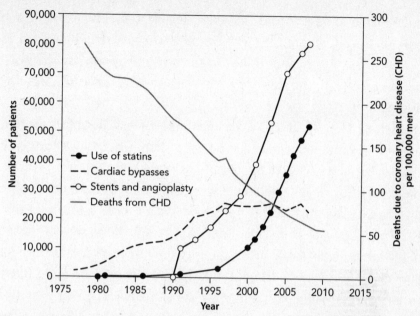

This chart shows the rise in the use of coronary bypass surgery, stents and other related procedures, and cholesterol-lowering drugs (statins) through time. Data are for the United Kingdom, but similar trajectories exist for the United States and elsewhere in the affluent world. Also shown is the declining number of deaths due to heart disease through time due to changes in the prescription of drugs, the use of new procedures, and, to a lesser extent, diet.

(that is, drugs and lifestyle interventions). The review combined the results of all studies in which treatments (stent versus non-stent) had been randomized.[4] Collectively, these studies included 7,229 patients, about half of whom were given stents and medicine and half of whom were given medicine alone. It is the biggest such analysis to date. On average, the patients in the studies were followed for a little over four years. Given the cost and sophistication of stents, one would expect those individuals who received them to be less likely to require another heart intervention, less likely to have a heart attack, and less likely to die. Statins and other medicines involve taking pills; stents involve having a device run up your blood vessels and into your coronary artery, where it will remain. Surely, stents are much better than the use of drugs alone. It scarcely seems necessary to make the comparison. Stents open up the artery and hold it open; that this is progress can hardly be doubted.

It should be. With stable coronary atherosclerosis, stents are not, on average, any better at reducing the risk of heart attacks or death than treatments with drugs and behavioral changes are. Overall, the results for drugs and stents were the same. Where there were differences, the medical therapy alone (drugs and diet) fared better. Patients who were prescribed medical therapy alone were no more likely to require subsequent intervention or surgery than stent patients. They were no more likely to die. They were no more likely to throw a clot. The only difference appears to be in terms of symptoms. With medical therapy alone, patients were less likely to have resolution of their symptoms, including their chest pain. In other words, the Canadians very clearly have it right (depending on how much we weight the treatment of symptoms). Despite the stents' long history. Despite the brilliant men who pioneered them. Despite the billions of dollars made placing them. Stents show no evidence of being any better as a treatment for stable angina than just taking the right drugs and living a lifestyle along the lines of the one proposed by Ancel Keys, including exercise.

The results of two studies performed in the 1980s and 1990s conflict with this new finding, but both do so superficially and in expected ways. In the 1980s and 1990s, the comparison that could be made in studies was between angioplasty and treatment with aspirin and ACE inhibitors (but no statins). In this comparison, angioplasty is better. But when stents are compared to treatment that includes Endo's statins, stents do not perform better. At a cost of more than thirty thousand dollars each, they perform extravagantly the same. Death rates were about 9 percent in both groups, and among those who did not die, rates of nonfatal heart attacks were slightly higher for those who had stents than for those who just took medicine.

The choice of solutions in different regions seems to have as much to do with money and local culture as with success. In the United States, most hospitals have gone to fee-for-service systems. Hospitals and doctors are paid more if they perform procedures, which is especially (economically) beneficial if those procedures are relatively easy. The stent falls precisely into this category. Many, many doctors can now place stents. They do so easily, at least when compared to the challenges of a coronary bypass. These stents have made hospitals huge money, but the stents collectively cost patients and health-insurance companies billions of dollars a year in the United States alone, most of which goes to the doctors, the hospitals, and the companies that produce stents. In some hospitals, half of all revenue comes from inserting stents. As one doctor put it in an interview with the *New York Times*, "When you put in a stent, everyone is happy—the hospital is making more money, the doctor is making more money."

But how could this be? Stents do seem to open up clogged arteries. Bypass surgeries create new arteries. Both approaches match up with our intuition about how to clear a pipe. Here there exists an important subtlety revealed only recently, a subtlety in the biology of atherosclerosis, a subtlety that reminds us of the many ways in which arteries are far more than simple pipes. Well into the 1980s, physicians and researchers assumed that the clogging of arteries was progressive. Cholesterol and immune cells accumulated in artery walls until the ever-thickening walls simply touched. This is in line with da Vinci's speculation but also with the thinking of nearly everyone working on the heart over the past hundred years. Physicians and researchers assumed atherosclerosis, like the accumulation of sediment in the bottom of a river, was slow and progressive. But in the 1990s, it began to become clear that this model was, just maybe, not quite right.

In the 1990s a series of studies showed that arteries clog not because of progressive narrowing alone, but because of a combination of narrowing, atherosclerosis due to plaques, and ruptures of those plaques. When plaques rupture—break open—they spill their contents into the arteries. Those contents—a messy mix of cholesterol, immune cells, and triglycerides—then trigger blood to clot. It is this clotting that clogs the arteries; it is the clotting that deals the final blow. While both stents and bypass surgeries open the coronary arteries, they do not clear the plaques out of the rest of the arteries, and so when plaques in arteries anywhere in the body blow, the mess that results can travel to the coronaries and clog them. What is more, while bypasses and stents often target big plaques, recent work suggests that it is actually intermediate plaques that are most likely to rupture.

In light of this model of artery clogging, the thrombosis model, the limited and somewhat idiosyncratic success of bypasses and stents make sense. The arteries are not just pipes that clog; they are living tubes subject to phenomena that are complex and poorly understood. This model also suggests some of the reasons that aspirin and beta blockers (and even statins) work. Statins work because, just as Endo thought, they reduce cholesterol levels in the blood, but more specifically, they lead to the removal, by the body, of cholesterol in existing plaques. Aspirin works by decreasing blood clotting, which makes clogs due to thrombosis less likely (in other words, you take aspirin if you have clogged arteries because it will help prevent a heart attack on the one day a plaque does break open). Finally, beta blockers help reduce blood pressure, which reduces the force of blood on plaques, which reduces the chances they will burst. All of these drugs are different from stents in that they reduce atherosclerosis, the threat of thrombosis, or both, and in doing so, they systematically decrease the risk of the arteries clogging so severely a heart attack occurs. Remarkably, after a century of racing into the heart, a century of trying to mend or even replace that vital pump, we are only just beginning to understand which elements of what has been achieved have actually been successful.

Other successes have come in terms of prevention of heart disease and treatment of its consequences, though the relative significance of different preventive measures is hard to judge. Most research findings are similar to the *Archives of Internal Medicine* study results in that they highlight the importance of drugs that reduce blood pressure (such as beta blockers), drugs that reduce blood coagulation (such as aspirin), and drugs that reduce levels of cholesterol and inflammation (statins). There's also another important factor in heart-disease prevention: a reduction in smoking.

There is no Ancel Keys in the story of smoking and heart disease, no singular advocate or discoverer. Instead, in the 1970s, it began to be noticed that smokers were at greater risk of heart disease. There was never any controversy as to the existence of such a link. The debate revolved around why it existed. The reasons appear to be multiple. Smoking increases the risk of heart disease by constricting arteries and increasing the incidence of clot formation in blood. It also increases inflammation by introducing small particles into the lungs to which the immune system responds as though they were living, foreign, and dangerous. It does all of this in both those who smoke and those who experience smoke secondhand.

Smoking is interesting in terms of its effects on our bodies. But it is also interesting in the context of how a decrease in smoking has been engineered. Public-service announcements proclaimed the ills of smoking, just as Keys proclaimed the ills of fat, but such proclamations did not have much influence. What did have an effect were laws, regulations, and taxes. As a result of these tools, tools of policy, it has become more difficult to smoke, more costly, and more unpleasant. In restaurants and other public places in most states, smokers must now cluster on back steps, blowing smoke in one another's unhappy faces as rain falls on them and the cold or hot wind blows. What was once a glamorous activity is now associated with social, legal, and financial marginalization. As a result, the incidence of heart disease (and lung cancer) has declined in lockstep with reductions in the number of people who smoke and the number of cigarettes smoked by those who do.

Changes in policies surrounding smoking improved the health of not only smokers but also those who didn't smoke. A recent report by the Institute of Medicine found that in countries as diverse as Italy, Canada, and the United States, reductions in smoking led to decreases in heart-disease risk. The extent of the reduction was never less than 6 percent and was as high as 45 percent in some regions. Even the most modest of these impacts, 6 percent, is more of a decrease than is seen when people on modern Western diets switch to Mediterranean diets.

The success of taxes and laws in discouraging smoking offers many lessons. One is that if you want to reduce negative elements in people's lifestyles, you must do more than just urge people to change. It is not personal choice that alters the healthfulness of behavior in the biggest ways; it is the landscapes society engineers, be they legal, political, or physical. One of the best examples of how powerful and immediate such changes can be comes from an effect on the heart that, like secondhand smoke, is difficult for any individual to control on his or her own: urban pollution.

Like smoking, air pollution causes increases in the rates of heart disease and heart attacks. It is the smallest particles of pollutants[5] (2.5 micrometers in diameter and less) that cause health problems. They affect our lungs, these particles, but they have bigger impacts on our hearts. Like cigarette smoke, they cause arteries to constrict (and hence blood pressure to rise). Like cigarette smoke, they cause clots to form. And like cigarette smoke, these particles trigger inflammation.

Recent studies suggest that reducing the concentration of these particles could reduce death due to heart attacks by about 12 percent. Before the Olympics were held in Beijing, Chinese officials cleaned the air over a period of months leading up to the event by restricting the use of cars in the city and decreasing the activity of power plants outside the city. When they did, the rates of heart-disease-associated deaths fell. After the Olympics, the power plants started back up and the cars once more took to the roads. Deaths due to heart attacks increased again. Apart from quitting smoking, getting rid of pollution can contribute more to reducing the risks of heart disease than nearly all of the other preventive measures that have been discussed so far. At least in theory, pollution, like smoking, can be controlled, and by its very nature, it is a public phenomenon. Short of moving somewhere with clean air or staying inside, no individual can alter his or her own fate with regard to pollution, but laws can affect pollution, as can the decisions that individual cities make. So can urban trees and, it turns out, beetles.

In 2002, the first emerald ash borer (*Agrilus planipennis*), a small, green, jewel-like beetle, was found in Detroit, Michigan. Soon there were two, then two thousand, then hundreds of trillions. Yes, hundreds of trillions. The borers are native to Asia and eastern Russia and are intruders in Michigan; they spread in shipped wood. In Asia, they are innocuous, but in North America, when they find ash trees, they kill them, slowly but inevitably. Just why these beetles are so much more deadly in North America than in their native lands is unclear, but it relates in part to the immune response of the trees they attack. The North American trees overreact to the beetles; it is this overreaction that leads to the trees' death.

Once established in Michigan, the beetles spread unchecked across the state and into adjacent states, killing essentially every ash tree in their path. Ash trees are often planted as street trees, and so the effects were greatest in neighborhoods. Suburbs that had once been shady warmed with sun. More than a hundred million trees have died so far (roughly one for every adult in the United States), and as the beetle spreads, even more seem doomed.

The death of so many trees is tragic. It affects the birds, the bees, and all the other species that depend on trees. But when Dr. Geoffrey Donovan, a scientist who works for the United States Forest Service in Portland, Oregon, thought about these trees, he worried about humans. Donovan had spent most of his career measuring the benefits and costs of trees. Trees affect property values. They can help reduce crime and increase the time adults and children spend outdoors. Trees also, it's been suggested, benefit human well-being and health, particularly cardiovascular health. In one study in Tokyo, for example, seniors who lived nearer to green space tended to live longer.[6] This could be due to the beneficial effects of trees on pollutants (trees pull many pollutants out of the air), though it could also be due to other benefits of green spaces.[7]

A study in which people's exposure to trees was altered while everything else in their lives remained the same would be compelling. Donovan realized that the emerald ash borer's invasion represented just such an alteration, an accidental experiment replicated in tree after tree across the Midwest. Donovan speculated that when the ash borer killed trees, it might also be killing humans by increasing the pollution around them in cities (which increased risk of heart disease and other cardiovascular problems). In theory, this was conceivable, but theory predicts many features of the world that never come to be. Yet Donovan had that great arbiter of theory: an empirical test.

Donovan and his colleagues compared the change in the number of ash trees in a region to the changes in health in those same regions. What theory predicted was that in those regions where more trees died, health metrics, particularly those associated with cardiovascular disease, should worsen. Making this comparison took time. The data sets on human health and tree deaths had to be combined. Analyses had to be worked out. Treacherous analytical bridges needed to be crossed. But eventually, Donovan had the results, and they were even more striking than the theory had predicted. Where ash trees died, cardiovascular diseases and deaths due to cardiovascular diseases became more common. Between 2002 and 2007, Donovan and his colleagues estimated, fifteen thousand more people appear to have died of heart disease in Michigan and eighteen other Midwestern states due to the death of the trees than would have died otherwise. The ash borer has proved more dangerous in the Rust Belt than any other animal (aside from humans).

The deaths associated with tree loss might not be exclusively associated with tree-mediated air pollution. Trees provide other health benefits. They reduce stress. They reduce temperatures (treed parts of cities can be as much as six degrees centigrade cooler than areas covered in cement). But pollution seems to be a big part of the story. The effects of the trees are greatest in more affluent areas, where trees tend to be planted in higher density, suggesting the possibility that losing trees is bad but losing more trees might be worse. Meanwhile, the beetle continues to spread. It seems likely to spread everywhere there are ash trees, from sea to shining sea. By some estimates, this tiny beetle could kill as many as 7.5 billion trees, one for each human on Earth, before it is done. With those trees gone, if Donovan and his colleagues are right (and if the effect of the loss of a tree is the same regardless of where it occurs), a tiny beetle might have killed as many as a million people.

If there is good news in this story, and I think there is, it is that if the death of trees can take lives, then planting trees should save them. While it has proved hard to help people change their diets and behaviors in ways that affect the health of their cardiovascular systems, it is easy to plant trees. It is easy to imagine new cities, cities in which policies make us less likely to smoke and in which trees can be gardened for our health.

*   *   *

The idea of improving our health by planting trees is simple and primitive, particularly when compared to the sparkling innovation of artificial hearts or transplants. It is, at its root, a public-health measure, and public health is rarely sexy, even though it contributes far more to people's health and well-being than medicine does. Public health brings us the power of sanitation, vaccination, and the monitoring of the spread of pathogens to save lives, billions of lives. Public health, as a field, also invites us to do something else, namely, to consider how we might improve the quality of life of the average person in a region, in a country, or even on Earth.

When treating an individual patient, a doctor (at least in the United States) is expected to do whatever can be done to prolong life. Not necessarily good life, or quality life, or one life over another. Just life. But from the perspective of society at large, from the perspective of public health, the goal is necessarily different. The goal is instead to increase the longevity or quality of the average life. Naturally, we tend to care most about our own health, the health of our family members, and the health of the people where we live. But let's imagine for a second that instead of extending our own lives, the goal was to extend the average life. What would we do?

If we were considering the entire world, the first step would not be dealing with heart disease. It would be treating or preventing malaria, HIV, and other pathogen-borne diseases so that the average child could live long enough to worry about heart disease. It would be ensuring that everyone has clean water so that deaths due to diarrhea could be prevented. For much of the world's people, heart disease remains a disease of the wealthy that affects people in relatively old age. May your children live long enough to worry about heart disease.

But that is globally. What about within the United States or within a particular city? Right now, part of America has experienced the benefits of the past fifty years of heart research, and life expectancy for them has increased. But many of the country's citizens haven't seen these benefits. For large segments of the American public, life expectancies are what they were in 1950, as are rates of heart-disease death. In the United States, genes can be predictive of heart disease, but better predictors, all other things being equal, are money and education. The poor and poorly educated have essentially missed out on the benefits of the past fifty years of medical science when it comes to cardiovascular problems. Individuals of lower socioeconomic status are at a 50 percent increased risk of developing heart disease than those of higher socioeconomic status. This is true even after risk factors such as diet, obesity, and smoking are taken into account, and it's due to disparities in health care and access.

This discrepancy appears to be increasing rather than decreasing. If we wanted to make major advances in the life expectancies of individuals in the United States, we could invent new technologies that eke a few more years out of the human body. Or we could plant policies, trees, and equitably distributed health care. The key is not necessarily to plant it with greater innovation, just with greater equity, so that everyone might live a longer and greater life. If you are already among those fortunate enough to have good health care, this won't improve your fate, but it will have a large effect on the average person's fate, the average human condition, the condition of those many who are not so fortunate. Today, those who are less fortunate with regard to heart care are the poor, whether the poor in wealthy countries or the poor in the impoverished nations of the world. But until relatively recently, the category of those who were not so fortunate, who didn't experience the benefits of the past decades of progress, included children, children born with congenital deformities and abandoned as beyond repair, children with broken hearts.

14

The Book of Broken Hearts

On November 29, 1944, Eileen Saxon, a fifteen-month-old girl, was on her back on a steel table at the Johns Hopkins Hospital. Inside her small body, her heart was doing something unusual, though just what, no one could say for sure. Her skin was blue. She was alive. Those who stood over her, including a chief surgeon and a cardiologist, thought she had blue-baby syndrome, in which the blood pushed through the body by the heart is low in oxygen. It is a common congenital disease, often caused by a suite of deformities collectively termed the tetralogy of Fallot, named for French physician Étienne-Louis Arthur Fallot. These four ("tetra") deformities are ancient, predating the origins of humans or even primates. They have been found to occur in every mammal species that has been studied well enough for us to detect cardiac abnormalities. The malformation has occurred billions of times, but prior to 1944, not once among those instances had it been cured. Each time, the baby—be it gorilla, squirrel, or little girl—eventually died, literally suffocated. Now, here was another blue child, but her situation was different. Gathered around her in their ghostly gowns and masks were a dozen or so people, at least three of whom believed she might be saved.

In 1784, William Hunter at St. George's Medical School in London first described the causes of the blue-baby syndrome accurately and in detail: "(1) the passage from the right ventricle into the pulmonary artery, which should have admitted a finger, was not so wide as a goose quill; and (2) there was a hole in the partition of the two ventricles [a ventricular septal defect], large enough to pass the thumb from one to the other. (3) The greatest part of the blood in the right ventricle was driven with that of the left ventricle into the aorta, or great artery, and so lost all the advantage which it ought to have had from breathing." Also, (4) the right ventricle becomes hypertrophied, enlarged, because it must work harder to get blood through the narrowed pulmonary artery and into the lungs. These were the syndrome's four problems, which always seemed to occur together and which led to there being less oxygen in the body than was needed. Many cases of this disorder occurred in humans, about 1 in 3,600 births around the world, but to Eileen's parents, there was only one case that mattered.

The chief surgeon was Dr. Alfred Blalock, a surgeon in the children's cardiac clinic at Johns Hopkins. Blalock had gone to medical school at Vanderbilt (as the school's first student); he would do the cutting and sewing. Beside him, standing on a stool and whispering in his ear, was Vivien T. Thomas. Thomas was the one with the surgical skills, a genius with his hands. Thomas had performed all but one of the practice surgeries necessary to prepare to work on Eileen. In dogs, he created and then repaired problems like those seen in the tetralogy of Fallot. He could predictably save dogs. Based on his skills alone, Thomas should have been the one to do the surgery. But Thomas had grown up poor and black in a time where both made access to good high schools, much less colleges, difficult. He started college but dropped out because of a lack of funds. His way into surgery would be through a back door. Blalock had hired him as a technician, but he proved to be a genius at surgery, even if not legally a surgeon.[1] He would guide Blalock, his collaborator, much as if, some would later say, Blalock were a marionette and Thomas were the puppeteer.

These were the two people closest to the patient, but behind them was Dr. Helen B. Taussig. Taussig had invented the surgery that was about to be attempted, though she was not a surgeon either. Her gender had prevented her from access to surgical training. This was an unusual team. Even Blalock, its most orthodox member, was a bit marginal. He was shadowed by self-doubt and thought himself a failure; he had had trouble getting a surgical residency. Vanderbilt Hospital was the only place that would have him. He was among that hospital's very first residents.

Blalock told Eileen's parents about the dangers of the surgery the doctors were about to embark on. Her parents knew enough to realize this was the only choice. No child with blue-baby syndrome had ever survived past the age of four. Of course, there was a chance that she did not have blue-baby syndrome, but no one could know until they saw her heart. Blalock did not give the parents the complex biographies of his team. Nor did he mention his own hesitation—his persistent and gnawing fear. For the first time in the 130-million-year history of tiny broken mammalian hearts, this baby, their baby, might recover from the tetralogy of Fallot. If the surgery worked, Eileen's skin would go from blue to pink. Her fingers would flush with life, and, as they did, Eileen would get a second chance.

The buildup to this moment began with Dr. Maude Abbott, Maudie to her friends. Abbott was born in St. Andrews, Quebec, where adversity was her native substrate. Her father abandoned the family. Her mother died of tuberculosis. A grandmother raised her and her sister. Abbott was forged out of hardship.

Abbott entered McGill University in 1886 and graduated valedictorian of her class, but McGill did not yet accept women to its medical school, and so she had to find somewhere else to go. This led her to Bishop's University in Lennoxville, Quebec, where she was the only female medical student in her class; although she was admitted, she was never quite accepted socially. Because of the persistent barriers to women in medicine, her drive, ferocious as it might be, could not push her straight ahead, and so she would go

Maude Abbott pretending to read. *(Courtesy of Harris & Ewing/McGill University Archives, PR023284)*

sideways when she needed to in order to keep moving. Once she obtained her medical degree (again with the highest honors), Abbott opened a clinic focused on women and children. But, in her own telling, she did not have the right kind of empathy to deal with children and suffering. She was better at autopsies; the dead demanded only her persistence. She gained some renown when she identified hemochromatosis (a disease due to too much iron in the blood) in one of the bodies she studied and wrote a paper on it. On the basis of this paper, George Adami, the chair of pathology at McGill, recommended her for a position as an assistant curator at the Medical Museum of McGill. She was offered the job in 1898, and she took it. There, she began the work that would lay the necessary foundation for Blalock, Taussig, and Thomas.

The museum charged Abbott with organizing the collection of body parts—organs and other pieces—left in disarray by her predecessor, William Osler. Osler was a brilliant pathologist, so dedicated to understanding the body's failings that he would drive a hundred miles to see a fresh cadaver.[2] He didn't care for medicine, for curing the living. His interest and his genius were in figuring out the cause of death, case by case, as in any modern autopsy. From each of the more than one thousand autopsies Osler performed in Montreal, there were parts arrayed in over a thousand jars. It was a liquid-y rogues' gallery of how human bodies go wrong—valuable, but not yet valued. Abbott considered these parts and sought to organize them the way a taxonomist might group birds, according to their similarities and differences, into basic kinds. Based on these parts, Abbott compiled the most comprehensive list in the world of the congenital defects of the heart, including the one leading to blue-baby syndrome, the tetralogy of Fallot. It became a gruesome traveling exhibit of broken hearts. It was first shown at the Graduate Fortnight in Cardiology at the New York Medical Society in 1931 and then again and again. It ultimately became a standard feature of medical-school classes at McGill. Then, in 1936, the same year she officially retired, she published the book that helped to frame the understanding of what can go wrong in a heart, *The Atlas of Congenital Heart Disease*, which is still in print today, by the American Heart Association. The atlas showed the diversity of the human heart's failings as recorded in one thousand congenital malformations, all of which, Abbott imagined, or at least hoped, might someday be remedied.[3] Abbott wanted to build on the book's success and write another book, this one a textbook on the heart. She received a grant from the Carnegie Foundation in 1940 for the project, but at the age of seventy-one, she died of a stroke before she could complete the work.

Many books have compiled the problems of the human body. Most have all the flair and seriousness of sideshow oddities. "Here is the woman with a hole in her heart. Come see the man with an extra ventricle!" Abbott's book was different. It spoke authoritatively and comprehensively to what congenital deformities looked like in autopsies. Yet it still said very little of their signs and symptoms in the living, much less how to treat them. Few followed up on it other than to add new instances of problems of the heart, new jars on the shelf. Then came Dr. Helen Brooke Taussig.

Like Abbott, Taussig seemed to find an unusual level of adversity en route to her successes. Her mother, like Abbott's, died when she was young. She struggled through high school with dyslexia. Then her hearing, the very sense she would ultimately use most to try to diagnose problems, started to mysteriously disappear when she was in her early thirties. Add to this her gender. Because she was a woman, she was not able to get an internship in surgery, so she accepted one in pediatrics. Then luck came her way. In 1928, one of her mentors, Edwards Park, opened a pediatric cardiology center at Johns Hopkins, the first of its kind and the only one for decades to come. Park thought children—chronically ill children in particular—needed better care; they were suffering and ignored.[4] He sought out Taussig

William Osler standing beside a live patient in front of a captive audience. This surgical theater, like modern ones, is a direct descendant of both the coliseums in which gladiators fought and the stages on which Galen performed. *(Courtesy of the Osler Library of the History of Medicine, McGill University)*

to lead the implementation of this vision, a radical vision that ill-fated youths might be saved. There, Taussig saw hundreds of desperate children suffering from disorders of the heart.

It was becoming clear at the time that rheumatic fever, the result of an overreactive immune response to an infection by *Streptococcus* pathogens (the bearers of strep throat), was the leading cause of heart failure in children. The reaction of the immune system to a particular group of *Streptococcus* pathogens (group A) leads to a thickening of the heart valves, particularly the mitral valve, and a restriction in the valve's movement. When the restriction is complete, or nearly so, heart failure results. Rheumatic fever and, with it, rheumatic heart disease seemed like an obvious target for research, as early medicines had already emerged to treat the pathogens (sulfonamide drugs, long the most effective treatment against bacteria), if not necessarily the immune response. Also, the sense that new antibiotics were on the horizon seemed to be in the air. Taussig found joy and fulfillment in working with patients with rheumatic fever (which was a leading killer of children in the United States then, as it is today in much of the rest of the world), but others in the hospital objected when she treated these children. She was "stealing their cases," and so she focused on the other heart problems, the harder ones, the congenital ones, the ones with absolutely no hope; this was her allotment in life, an allotment of death.

In response to such challenges, Taussig was "aggressive, defensive, combative, sometimes triumphant, and often defeated. She suffered."[5] Her friends told her she had chosen (though that hardly seems the right word) the wrong field. As historian Laura Malloy put it, "The prevailing viewpoint was that even if congenital malformations could be accurately diagnosed, nothing could be done about them."[6] But Taussig believed in what might someday be possible. Unlike Abbott, her empathy knew no bounds. She felt she had to work on what she had been given, these children who were beyond cure and whom no one else would even try to help—children like Eileen.[7]

A piece of rheumatic heart from the Osler collection curated by Maude Abbott, one of her many "jars" in which the variety of our dysfunction could be made obvious. (*Courtesy of the Maude Abbott Medical Museum*)

As Taussig began to work on the hearts of children, she carefully read Abbott's book. She even saw her display and met her. Later, the two developed a relationship that would have been described as mentor and mentee if the two women were not both so strong-willed. Their relationship, like much about each of their lives, defied simple characterization. In looking at what Abbott had compiled and in talking with her, Taussig was struck by the diversity and frequency of human maladies. Roughly one in 125 children is born with a congenital heart deformity, and this is not even accounting for those stillborn babies whose hearts never fully developed; heart deformities are the most common congenital problems. She was also struck by something else. Taussig noted that although there did not seem to be that many kinds of congenital heart problems, they repeated. One could imagine that, with all its genes and stages of development, the heart might develop an infinite variety of problems, one for each of the combinations of mutant cells, but Taussig saw something different, something hinted at by Abbott's work. To her, the same problems seemed to recur, as if the problems themselves had some predictable genetic origin, not in randomness but in evolution. At the time, congenital defects were thought to be due exclusively to exposures to dangerous environments and substances that affected the genes, broke them. To Abbott, this, from the very beginning, seemed wrong, or at least only partially right. The stories were not an infinite number of unique kinds of damage, but rather a smaller number of repeated forms. Abbott had hoped for cures for the congenital disorders she found, but Taussig was determined to do more than hope. She would try to find cures, ideally for the most common forms.

To do so, she had to figure out how to diagnose more of these problems in living children. MRIs, catheterization, and angiograms had not yet been invented. Forssmann had not yet probed his heart. The insides of the body were invisible unless one cut in. When Taussig was given her job in 1930, her equipment for studying the heart consisted exclusively of an EKG, the device used to track the electrical rhythms of the heart. If she was to learn anything more than what the EKG could tell her, she would have to rely on her own senses. So, she looked and listened. She described the endeavor as being like a crossword puzzle in which the answer was a disease and the clues were those garnered from listening, from blood pressure cuffs, and, later in her career, from the hazy images on fluoroscopes. Fluoroscopes are simply x-rays observed in real time so that the body's internal movements can be captured. Each puzzle's tragic conclusion, though, was that Taussig would know if she had answered correctly only after her patient died. Then she could look inside the tiny body and see if she was right. This was awful, soul-crushing work. There were almost never happy endings—only tiny tragedies from which she could learn incrementally more.

Taussig felt she had a duty to chronicle heart diseases so that, if someone ever discovered a way to cure one, the curable disease might be diagnosed in other patients. By 1944, building on Abbott's work, Taussig had compiled a relatively complete catalog of congenital heart diseases, including the signs of each—the tracks left on the body by the disease's effects on the heart. Her book became the first standard book in pediatric cardiology. As one of her colleagues, Carleton Chapman, would later say, "That book made all the difference. It brought congenital heart disease out of the fairy land" and into the realm of treatment. Some defects, she argued in her book, could always be predicted based on symptoms. Others could not. The cause of the blue skin of blue-baby syndrome could be predicted; it was nearly always due to the four deformities of the tetralogy of Fallot.

As of the publication of her book in 1944, Taussig could reliably distinguish the tetralogy of Fallot from other diseases that cause the skin to turn blue, but there was still nothing she could do once she made the diagnosis. She braced the parents and waited for the body. It was about this time when she heard some discussion about surgically sealing holes in the heart. She wondered about the possibility of doing the reverse, opening more holes, "arteries" of some sort that could lead from the heart to the lungs. These were the exciting boom years of heart surgery—the heart-lung machine was on the horizon—and Taussig was attuned to the possibility that some of the new approaches in surgery might be used to save the children dying all around her.

Most of the heart surgeons racing toward progress focused on adults; adult hearts were bigger and easier to operate on, even if the gain would typically be years rather than, as might be the case in children, whole lives. With a few prominent exceptions, children were operated on only in special cases, when a new technique and a failing child happened to coincide. Taussig was not waiting for coincidence. She wanted to operate on tetralogy sufferers; she just needed to find someone to actually perform the surgery—someone who had the knife skills and could figure out what exactly the surgery should be.

Taussig traveled to Harvard to consult with a surgeon who had done an operation in which he had closed the ductus arteriosus in the heart of a young patient. The ductus arteriosus is a small passage from the vessel carrying blood to the lungs, the pulmonary artery, to the vessel carrying blood from the left side of the heart to the rest of the body, the aorta; it is open in the fetus and helps to shunt blood away from the lungs (which serve no purpose in the fetus). But in some children, the ductus fails to close, and oxygenated blood coming from the lungs pours back into the pulmonary artery (and hence goes through the lungs again). In 1938, Robert Gross, a resident at Children's Hospital in Boston, had performed a surgery in which he closed a ductus.[8] Gross declined to help Taussig, accusing her of idiocy for even asking (which redoubled her resolve). Taussig returned to Johns Hopkins, where she persisted with her idea, so much so that other colleagues blamed her stubbornness on her deafness. Perhaps she literally didn't hear the word *no*. It would take another two years before she found someone willing to try her new procedure.

The man who would do it was Alfred Blalock. When he joined

Johns Hopkins, it seemed like he had been sent by the Fates. He had the skills to try a surgery, and, inspired by Taussig, he had an idea about just which one. He could, he thought, take one of the arteries traveling from the heart to the body and reroute it to the lungs. But before he tried, he wanted to perfect the procedure on animals. For that, he turned to his assistant, Vivien Thomas, who developed the procedure, and then, over three long years, repeated it in dog after dog until it seemed just right. It was after these surgeries, after Taussig's two years of finding no help, and then after three years of waiting for Thomas to perfect things, that Taussig, Thomas, and Blalock stepped into the room with the little girl, Eileen.

Eileen's parents wanted their daughter to live. The surgical team wanted the hundreds of thousands of blue daughters and sons to live. But a dog heart is not a human heart, and so the truth was that the team had no real idea whether their new procedure would work.

On the operating table, Eileen Saxon was too little to want or even really understand. She suffered but had never known anything different. She stared up at the lights and then fell asleep as the ether kicked in. Standing above Eileen's body, Blalock cut a smile-shaped incision, beginning below her right armpit into her chest and going up between her third and fourth ribs, before pushing apart her ribs. This alone was a challenge. She was so small that Thomas had had to devise not just one but several new tools and approaches to fit her body and the circumstances. Once her heart was exposed, he could clearly see it beating. The tiny heart was all wrong. Fortunately, thanks to Taussig's work, it was precisely the kind of all wrong they had anticipated. Blalock clamped both ends of her pulmonary artery and then did the same for the subclavian artery traveling out toward the body. In the former, he cut a small hole into which he inserted the latter. He stitched the two arteries together with beaded Chinese silk. Suddenly, two riverbeds were joined, and twice as much blood flowed to Eileen's lungs—hopefully enough to move more oxygen through her blue body. He then used the silk thread to reconnect the ribs, applied sulfonamides to the body cavity, and, layer after layer, stitched the tissues back the way they had been. After the surgery, when the three doctors stepped back, Eileen's heart kept beating, and, as it did, her body began to turn from blue to pink. As Eileen's mother would later say, "When I saw Eileen for the first time, it was like a miracle....I was beside myself with happiness." So were Taussig, Blalock, and Thomas. The surgery had worked.

Tragically, Eileen died three months later of other complications of her heart's complex deformity. The surgery had prolonged her life, but not by nearly enough. Technically, though, the surgery did what it was supposed to do, and so it was tried again. The second surgery went well but again proved to be a relatively short-lived fix. It was on the third try that everything worked out the way everyone had hoped. The patient was, as Taussig commented, "an utterly miserable, small six-year-old boy...no longer able to walk." Blalock performed the surgery on him, with the same drama as the first—the same anxious parents, the same trembling child. The boy turned a lovely color, with "lovely normal pink lips."[9] He woke up and looked at Blalock. He blinked his eyes and said, "Is the operation over? May I get up now?" He could. He got up in the next days and weeks that followed and became a happy, active child; he went on to have a full life—one given to him by Taussig's idea, Blalock's hands, Thomas's practice, and Eileen's tragedy.

Word of the surgery's success spread. Just two years later, the procedure had been performed hundreds more times.[10] One British cardiac surgeon, Sir Russell, called the operation "so outstanding that it altered the whole approach to cardiology."[11] Blalock had become famous, though at a time when fame for physicians was still regarded as unethical in the United States, just as it had been in Germany for Werner Forssmann and as it would later be, to a lesser extent, for John Gibbon. This fame simultaneously delighted and worried Blalock so much that he attempted to resign. Thankfully, his colleagues convinced him to continue working. As a result, thousands were saved. Children were brought and sent by parents to Johns Hopkins so that they might be born again with mended hearts. Hundreds of parents wrote Johns Hopkins, begging for their children to have the surgery, which would prove to be successful in roughly eight out of ten cases. Today, 90 percent of babies born with the tetralogy of Fallot will live out normal lives. Hundreds of thousands of adults walking among us are alive thanks to the team at Johns Hopkins.

Taussig, Blalock, and Thomas would all become known for the tetralogy surgery. In every description of the events that led up to the new procedure, different emphasis tends to be put on the contribution of each of the three individuals (the approach is often called the Blalock-Taussig procedure, sadly omitting Thomas entirely), but the truth is that they were a team made up of individuals outside the mainstream of heart surgery who became great in the context of one another. Other surgeons would try to improve on the surgery (and name the "improved" versions after themselves), but the version that Taussig, Blalock, and Thomas pioneered persisted for years. It was replaced only when more elaborate procedures (in particular, the ventricular-septal defect patch repair and artificial shunt) were made possible thanks to Gibbon's heart-lung machine.

For Taussig personally, the surgery was one success among many for which she laid the groundwork. She studied children's hearts when everyone else ignored them. She listened to children, reading lips when she had to. She valued diagnosis when others thought it useless, and she was willing to do the tedious work. This work acquainted her too well with the death of children, but because of it, she made possible the greatest successes in heart surgery—the tetralogy surgery, but more than that, an entire array of interventions so successful that if a child is born with a congenital heart defect today, he or she now stands an 80 percent chance of having a normal life span. Taussig's success inspired others to establish pediatric cardiology wards all over the country. She helped to fund these centers through an appeal to the National Institutes of Health and the Children's Bureau, and she trained the next generation of pediatric cardiologists and surgeons, nearly one hundred in total, many of whom were women.[12] Taussig persisted as a scientist,[13] as a leader (she would become the first female president of the American Heart Association), and as a mentor, but surely her most conspicuous legacy is her living one, the adults whom she helped heal as children.

Most heart procedures gain patients a few years, at most a few decades; Taussig's legacy has given whole lives. If there is a pinnacle to the story of the heart and our attempts to understand it, this is it: the hundreds of thousands of people born with heart defects who walk around us every day, looking no different, feeling no different— who are simply, miraculously, alive.

But this is not the end of Taussig's story. She did not, as one biographer put it, go gentle into the twilight. She did not know how. A 1968 painting by James Wyeth shows her at the age of seventy, white-haired, cloaked in a dark dress, and lit from the front as if traveling into the glow of yet another new discovery. Even after she moved into a retirement facility, she continued to work as a Thomas M. Rivers Research Fellow at the University of Delaware. Treatment of congenital disorders had advanced so much by then that most children born with congenital heart disorders survived. She saw them walking around her when she went out, individuals who would not have been alive but for her determination. With a life of successes at her back, Taussig could have enjoyed the fruits of her labors. Instead, she refocused. She studied adults who, as children, had been treated for congenital heart problems; she wanted to know their long-term fate. But she also became interested in the origin of the problems she had studied her whole life. She was convinced congenital heart problems could be understood in the context of evolution.

15

The Evolution of Broken Hearts

Nothing in biology makes sense except in the light of evolution.
— THEODOSIUS DOBZHANSKY

In 1984, Helen Taussig moved to a retirement home. At night she slept in a building filled with other retirees, men and women living through the last phases of their lives. During the day she drove herself to the local Delaware Museum of Natural History. There, she pulled out birds that had been sent to her, plucked the feathers from their bodies, and cut into their tiny chests. One might be a warbler. Another a starling. Looking just beneath the skin, she found hearts, nearly all of which were ordinary bird hearts, four-chambered, perfect, and fascinating. She did ten a day or so before disposing of the feathers and writing down what she had found. Then she drove home, noticing en route the birds flying over her, birds on lines, birds even on her very own stoop. Mostly their hearts beat in rhythm, perfectly, moving blood to gut, beak, and wings. Sometimes, though, their hearts beat wrong. At least, that is what Helen Taussig desperately hoped.

In retirement, Helen Taussig had embarked on an entirely new career. In the late 1970s, she began to imagine that if she studied the hearts of nonhuman animals—mammals and then birds—she might understand why so many babies were born with broken hearts. At the time, the prevailing idea was still that congenital heart defects were caused by mutations that arose when mothers exposed their unborn children to dangerous environments and mutagens (also called teratogens). Parents were blamed for the difficulties and even deaths of their children. Taussig disagreed. She wanted to understand the causes of the disorders of the heart in the hope that doing so might clear parents of the responsibility they so often felt. Her approach was unusual: she sought to study the heart's deformities in light of evolution. For a physician, this was a novel endeavor. It required taking the detective's approach she had used for decades when dealing with individual cases and applying it to a much bigger story, one that had unfolded over hundreds of millions of years rather than just the nine months of development. Taussig began to read what was known of the evolutionary biology of human hearts and those of other animals. What she found fascinated her and inspired big, bold ideas.

Taussig approached the evolution of the heart with two guiding principles. The first was that an understanding of the evolution of the vertebrate heart would inform her (and others') understanding of the problems of human hearts. To evolutionary biologists, this was an old idea, but to clinicians it was new, radical even. The second was that if one could understand which animals suffered from which congenital deformities, it might be possible to determine if the deformities were genetic and, if so, when they had arisen. Any congenital deformities present in both birds and mammals, for instance, must relate to genes older than either group. Conversely, deformities present in only mammals or only birds had to be more recent phenomena.

In some ways, the difference between the culture of the surgeons and physicians with whom Taussig had spent her life and that of the evolutionary biologists whose field she was now entering could not have been greater, even in terms of something as simple as the number of practitioners. In the United States, a cardiology meeting such as the Transcatheter Cardiovascular Therapeutics Conference might attract as many as ten thousand cardiologists. In contrast, the main evolution conference in the United States has a big year when it attracts two thousand people, most of whom are students. Those two thousand faculty and students study not just human hearts or even humans but all of life. If we suppose there are about ten million species on Earth (no one really knows; I suspect a much larger number), this leaves about ten thousand species per evolutionary biologist. The cardiologists and other physicians study just one species, humans, and typically a single organ of that species. And whereas the physician wants to fix the problem, typically independent of big questions about why it arose, the evolutionary biologist is not concerned with problems. The evolutionary biologist, in other words, tends to focus on exactly the thing the physician is least focused on, and vice versa.

But there is one thing that the physician and the evolutionary biologist share: the mind-set of a kind of detective. As Taussig started to look at the birds and the millions of years of evolution they implied, the sweep of time was new, but the love of the mystery reminded her of every mystery in every body of every child who had ever come into her office.

In reading the papers of evolutionary biologists, Taussig was fascinated by the number of forms a functional heart could take, its beating diversity. Most vertebrates—those animals with backbones, like humans—do not have a four-chambered heart. The fish have two chambers, as do amphibians such as frogs, and turtles, snakes, and lizards have three; birds and mammals alone have four. But what was fascinating to Taussig about this state of affairs, other than the question of how all of these very different hearts work, was that the birds and mammals seemed to have come by their four-chambered hearts independently. Birds evolved from one group of four-legged reptiles; mammals from another. Their most recent common ancestor lived more than three hundred million years ago. That both mammal and bird hearts have four chambers is a consequence of the predictable efficiency gained by having four chambers.

Taussig knew that she was biting off more than she could chew with this project. She decided to focus on the simplest piece: comparing the heart deformities in birds and mammals. Were they the same? Initially, she imagined that she might be able to just look at what was already known of the bird deformities and compare that to what she had learned about humans and other mammals during her career. The challenge would prove greater than she had anticipated.

Few scholars had studied the congenital deformities of hearts in nonhuman animals. When Taussig searched for examples, she found ancient anecdotes but few modern data. The Greek scholar Theophrastus, for instance, noted that all of the partridges of Paphlagonia seemed to have double hearts. This was fascinating, but she needed much more, which is what led her to begin to dissect birds herself.

You might imagine that Taussig's friends and colleagues would have greeted anything she did in her later life, after all the accolades and successes, with enthusiasm. She had earned enthusiasm. But little was forthcoming. Her colleagues and friends don't seem to have appreciated the work she did in her final years; it would later occupy a brief pair of vague sentences in the key biographical paper about her. They gave her the begrudging respect of never telling her that they thought this evolution work she had begun to do—every day—seemed to be, at best, an eccentricity. They may have thought her a victim of her aging mind.

The task at hand, mad or not, was difficult. Helen Taussig, who spent her entire career studying and helping children, had to figure out, essentially on her own, how to pull the feathers off birds and dissect their tiny hearts. What was more, she was going to need to study a lot of birds. If one wants to study the evolution of, say, a beak, one needs to measure just a few beaks to know what they tend to be like in a particular species. But Taussig was not interested in the average condition; she was interested in those anomalous ones, and so she had to study many, many individual animals to measure each of those anomalous conditions even once. To discover deformities, she dissected the bodies and hearts of hundreds and then thousands of birds. She told colleagues to look for dead birds. She scoured the streets herself. She made friends with someone who traveled to radio and television towers where birds often died due to collisions.

Most of the birds she ended up with were tiny. Her hands, long used to the small bodies of children, had to grow accustomed to the even smaller forms of sparrows and starlings. While her friends enjoyed their retirements, she performed the most difficult dissections of her life. She drew what she saw. She saved interesting hearts. She took notes; she pondered. She tried, after a career among hearts, to make sense of it all. She could have gone on forever, dissecting more animals and records to see the diversity of ways that hearts can go wrong, but she felt age creeping up on her and leaping at her mind. She drafted a paper based on what she saw. Things were going well, but just to be on the safe side, she provided instructions to her friends of what they should do with her late-life work should anything happen to her.

By the time she drafted her bird article, Taussig had already produced the most comprehensive paper written at that time (and to date) on the congenital deformities of mammals. In her focal mammal, humans, she found that the same congenital deformities were common all around the world, independent of where or how people lived. This, on its own, suggested to her that the deformities were not strongly associated with the environment, with mutagens, for example. When she looked at those nonhuman mammals that had been studied well, they too showed the same congenital deformities. Dogs and sheep, for example, suffer the same congenital deformities as humans in the same relative frequencies. The question now was whether the same would be true of birds.

Taussig predicted that those congenital deformities that related to shared, ancient features of the heart should be the same in birds and mammals but that those deformities related to the new features of the heart, features that had evolved since the separation of birds and mammals, should be different. In both cases, she was right. In the paper she had been working on, Taussig showed that some kinds of heart deformities, particularly those that caused cyanotic (blue-baby) diseases, were common among many mammals but also among birds. These defects, shared across the mammals and birds, seemed to Taussig to reflect problems in development at least three hundred million years old (the time at which the lineage of birds and humans split). Other problems seemed unique to the birds; she never saw a bird with two hearts, but two-hearted birds seemed to have been convincingly documented by others. Then there were ventricular-septal defects in birds, in which there was a hole between the two ventricles. Superficially, these defects looked like those in mammals, but their details were consistently different.

Taussig took these observations and came up with a big new set of ideas about congenital heart diseases. She had explanations not only for the origins of congenital heart disease but also for the origins of the complexities of the heart in the first place. Then, on May 21, 1986, while driving her friend to a polling station near her home in Kennett Square, Pennsylvania, Taussig's car was hit by another car she failed to see coming. She died later the same day at the hospital, three days before her eighty-eighth birthday.[1] Taussig's friends honored her wishes and published her paper, despite their skepticism, two years later, in 1988, though they would preface it with caveat after caveat in doing so, even including in the introduction to the paper the line "[Taussig] had no intent to label this work a scientifically based research project." Five thousand birds is an awful lot of birds to dissect for someone who doesn't think she is doing science-based research. Ultimately, her friends were too quick to apologize, and they were wrong to have doubts. Taussig's paper was brilliant, science at its most creative. But no one ever discovered

Taussig's paper. It is referenced just once by another scientist (writing in Polish), except where it is mentioned in biographies of Taussig as an example of how she kept busy in her old age.

Fortunately, during the past ten years, the ideas that so captivated Taussig have begun to be explored independently by evolutionary biologists. Evolutionary biologists have spent the past hundred years or so trying to reconstruct the big evolutionary story of the heart, how it evolved, and how it functions under normal conditions. Now they have begun to consider, in light of this big evolutionary story, human heart problems. As they have, what they've found suggests that Taussig was onto something. Considering the evolution of the heart alters how we think about congenital heart disease but also how we think about that even more common plague, coronary artery disease.

The story of the evolution of the heart is embedded in every heart and how it works or fails to work. The heart and its problems, as evolutionary biologists tell it, begin not with birds and mammals but instead at some point prior to 550 million years ago, when the first multicellular creatures appear in our fossil records of the sea. Nutrients and gases could diffuse into single-celled creatures, but as organisms grew larger, some cells were inevitably on the inside of the creatures of which they were part. Those inside cells required plumbing[2] so that nutrients and gases could get to them. The first blood vessels emerged before the first heart.

Sponges (which are animals, though just barely) have the simplest circulation systems of any living organism. A sponge does not move. But it is filled with pipes through which the sea can be moved. Thin hairs in those pipes urge seawater along. The sea itself is part of the sponge's heart. As the water moves, the cells lining the sponge's network of pipes and tubes extract nutrients and oxygen from the sea and release waste.[3] The sponge's system may seem crude, but it works sufficiently well that sponges continue to thrive.

It is interesting for that reason alone, but also because evolutionary biologists suspect that the sponge's humble pipes are very similar to our own cardiovascular system's antecedent. Some of the human genes associated with veins and arteries are the same genes involved in producing the simple tubes of the sponge.

From this beginning, the descendants of the first sponge eventually evolved bodies big enough to require a pumping heart. In tens of millions of years — which to paleontologists is a very short stretch of time — many new and larger lineages evolved. In the few sites where good fossils of early multicellular creatures have been found, such as the Burgess Shale (in today's British Columbia), paleontologists have found a circus of strange multicellular life, a record of evolution as it explored different body plans. These early animals were more diverse in nearly every external feature than are modern animals, and one presumes the same was also true of their internal features. They may have tested out a variety of types of hearts. Many of the Burgess Shale species would disappear, extinct before they really got started, but several persisted, one generation to the next, until today, and in each one of those persistent forms — the ancestors of mollusks, worms, insects, and vertebrates — there is a primitive heart. Today, the genes for hearts in the descendants of these lineages are similar, suggesting they all derive from one ancient invention of the heart in the time of the Burgess Shale or slightly before. Your heart and the heart of a worm had the same beginnings.

In most lineages of animals, the heart stayed simple, nothing more than a squeeze box of spongy muscle, even in the earliest members of the vertebrates, the subphylum that includes humans. The first vertebrates were fishlike but not yet what we would recognize today as fish. In these animals, the heart's primary role appears to have been to move nutrients around the body. Nutrients were gathered by netlike gill structures and sent through the blood. The heart squeezed blood in both directions, fore and aft. When it did,

Colleen Farmer, a biologist at the University of Utah argues, it also squeezed oxygen-enriched blood from the skin into the nooks and crannies among the heart's spongy cells. (As the heart expanded after contraction, the oxygenated blood would have entered from the skinward side.) The very first role of the heart included supplying blood to itself.

With time, fish evolved to feed primarily with biting mouths, and when they did, the role of the gills became exclusively that of sifting oxygen from the sea and releasing carbon dioxide. Simultaneously, the skin ceased to be an important organ for obtaining oxygen. With the advent of a biting mouth, the heart of the fish became more complex. It evolved a two-chamber structure very much like what a human heart would look like if it had just a single atrium and a single ventricle. The atrium allowed more blood to collect so that when it was pumped, the pressure generated was higher. The blood then pumped around the body in one cycle: Heart-gills-body. Heart-gills-body. This means the blood returning to the early fish's heart was always low in oxygen because the other tissues of the body got to the oxygen first. The same is true in most modern fish. As a result, fish are subject to sudden death after vigorous swims, sudden death akin to a human heart attack in that it is due to a lack of oxygen to the heart.

Aside from making fish susceptible to sudden death from exertion (which they mostly appear to avoid through knowing their own limits), the fish heart is strikingly elegant. The human heart, like all mammal hearts, has two separate circuits—the left pumping blood out to the body, the right pumping it to the lungs for oxygen—while the fish heart gets the job done with one. In fact, fish hearts more than get the job done. By any real measure, fish have been more successful than mammals. There are many times more fish species than mammals species. That their system works begs the question of why our system became so much more complicated. The explanation for our complexity lurks just beneath the waves.

*   *   *

Lungfish are strange. They have gills, like other fish, but also, as is perhaps obvious from their name, lungs. To use their lungs, they reach up to the surface of the water and gulp air with their funny lips. They seem like primitive vestiges of something, an evolutionary bobble that has survived. But they are also a clue to understanding our hearts and their fallible complexity.

Lungfish were first discovered in 1837. A specimen was packed in clay and shipped to the British anatomist Richard Owen for study. Owen was well prepared to consider a new kind of fish. He had examined the bodies and bones of more species of fish than nearly anyone in history. He had an eye for subtle differences, a kind of visual intuition honed by thousands of hours of practice. But this damned fish confounded him. On the outside, it was clearly a fish, but as he peered inside its body, it seemed as if, in some fairy realm, the insides had been switched out and replaced with those of a snake or a frog.

Beginning with Owen, lungfish came to be viewed as interesting but rare anomalies. Yet, eventually, scientists realized there had been hundreds, perhaps thousands, of lungfish species. They were once a dominant life-form. As we now understand it, the story goes something like this: First, fish had gills to gather food. A subset of those fish evolved lungs in addition to their gills, primarily to gather oxygen, and then those fish, the lungfish, became rare. In light of this, the new questions become why and how lungfish became so successful and then why they became rare. Answering these questions would be of purely academic interest, except that the fish from which all terrestrial vertebrates, and their hearts, descended was a lungfish.

About 360 million years ago the first lungfish climbed onto land. Of course, there were many challenges to being on land. The lungfish had to evolve feet from its fins and deal with extra gravity, but the lungs made land and humankind possible; they allowed our first terrestrial ancestors to get enough oxygen to their hearts.

Although Darwin suggested lungfish evolved lungs from swim bladders (organs used by fish to stay buoyant), the opposite is true: swim bladders evolved from lungs. Lungs allowed fish to get more oxygen to their hearts in order to be more active in fleeing and chasing. The lungs were a benefit to the activity of the heart in the sea (just why lungfish became rare in the sea after conquering the land is not clear),[4] and they offered the same sort of benefit on land. That we have lungs to which our hearts pump blood is a quirk of our ancestors' attempts to be predators or avoid becoming prey.

Once these vertebrates were on land, an arms race began, an arms race that led to the major lineages of terrestrial vertebrates and was made possible through the evolution of the heart. Any descendant of the original lungfish that was more mobile than others could capture hard-to-catch prey or flee hard-to-escape predators. As a result, some lineages began to evolve hearts that were better at distributing oxygen, allowing more activity. It was a kind of evolutionary treadmill. Being more active—whether to catch or to flee from another animal—required more oxygen, which required a bigger and better heart, which in turn required more oxygen and food.

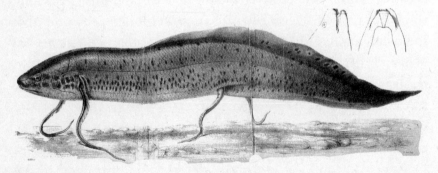

Richard Owen's drawing of the lungfish *Lepidosiren annectans* (now called *Protopterus annectans*), a species that is similar in many features, including both its lungs and its feetlike fins, to the ancestral lungfish from which all land vertebrates ultimately evolved. (*Courtesy of the Proceedings of the Zoological Society of London*)

The first new (and still extant) vertebrate lineage to evolve on land from lungfish was that of the amphibians. In amphibians, the two-chambered heart is sufficient as long as they move slowly and don't stray far from the water (where they can gather additional oxygen directly through their skin). The circulatory systems of amphibians are like those of lungfish. But this keeps amphibians bound to the water, tethered by their lunglike skin.[5]

The lineage of lizards, snakes, and turtles evolved a bigger, more efficient heart with two atria (like the human heart) and a partially divided ventricle. This new heart looked much more like a modern human heart than any that had come before, especially as it now had two circuits, one that ran blood from one side of the ventricle to the lungs and another that ran blood from the other side of the ventricle to the body. This heart allowed lizards and snakes to colonize the great first continent all the way to its inner reaches. This heart introduced a new problem, though: the blood coming back from the body through the right atrium was devoid of oxygen. Lizards, snakes, and turtles dealt (and deal) with this by having a hole between the nascent chambers of the ventricles so that oxygen-rich blood from the right side sloshes to the left. This sloshing works to nourish the heart muscle but is inefficient.[6]

Then, very recently, roughly around 180 million years ago for mammals and a little earlier for birds, warm-bloodedness evolved. Warm-bloodedness offers multiple advantages. It allows animals to be active all the time, even when it is relatively cold. Warm-bloodedness also helps to prevent colonization by many pathogens. The fungi that plague reptiles and amphibians, for example, mostly leave mammals and birds alone. Our bodies are too hot for mushrooms to grow in them. But these advantages come at a cost. Warm-bloodedness requires a constant supply of oxygen to cells all around the body so that they can metabolize and, in doing so, produce heat. As a result, warm-bloodedness requires the heart to be much more efficient (it also has to pump more frequently). Evolution dealt with this cost through the origin of a fully segmented ventricle. Blood no longer sloshed back and forth from side to side. There was no time for such inefficiency.[7] These ventricles evolved a full division with no room for leaks (creating a new problem in terms of oxygen supply to which we will soon turn). Birds and mammals both evolved this four-chamber system, independently. What was good for the archaeosaur (the first flying reptile) was also good for the first rat-size mammal running beneath the feet of dinosaurs.

All of this evolutionary context (with a few updates to reflect our modern understanding) that Taussig had begun to grasp led her to think that, in light of evolution, congenital heart diseases might make sense. Taussig noted that those rare congenital disorders of the valves or the left atrium or ventricle, all of which are as ancient as fish, seem to be found in essentially all vertebrates. The rarer deformities of the atria are found, typically, in those land animals that have added a second atrium. But the most common deformities all seem to relate to the right ventricle and those parts of the heart that formed with it. The right ventricle is the new one. The left ventricle corresponds to the original ventricle of the fish, amphibian, or lizard. The right evolved in mammals and birds. Taussig did not fully understand what this meant in terms of these congenital problems, but it seemed important. The major deformities arose with the origin of a second ventricle, but why? We now know.

During development, the heart, to a greater extent than any organ, recapitulates many of the changes it went through during evolution.[8] It starts as a tube. It goes through some loops and hoops, and then, for a time, resembles the heart of a fish, with one ventricle and one atrium, and then that of a lizard, with a single, large ventricle. If everything goes right, that big ventricle then divides partially during fetal life and closes fully just before birth. Recent research has revealed that the activity of a single gene, *Tbx5*, governs many aspects of the late stages of heart development in mammals and birds. In snakes, turtles, lizards, and amphibians, this gene and the genes it controls are expressed—the first step in going from the code of the gene to the physical product of the protein—across the entirety of the ventricle during development. Not in mammals and birds. In these animals, it is expressed in the left ventricle and then stops, abruptly. This pattern in expression tells other genes where to act. The placement of the abrupt stop determines where the wall between the ventricles is located. The evolution of the extra ventricle of both mammalian and bird hearts appears to have happened through changes in how this single gene is expressed. As Taussig anticipated, adding a new chamber was not so complicated after all. It mostly required a change in the heart's template. But this simplicity came with trade-offs.

It appears that in children with ventricular-septal defects, the expression of this gene goes wrong. In these children, the gene is expressed in the old way, the way it once was in our lizardlike ancestors. This leads to a lack of a wall between the ventricles, or a hole between the ventricles. Although it has not been well worked out yet, many other congenital disorders of the right ventricle, perhaps even including the tetralogy of Fallot, also relate to the expression of this gene. We still do not know what proportion of congenital heart diseases are genetic, but it is clearly higher than thought during Taussig's life. It may be very nearly all of them.

Taussig, despite her age, or perhaps because of her age and perspective, was right. Other scientists have only just begun to catch up to the ways in which she was right. Although individual studies confirm that most congenital heart defects are genetic, and although detailed genetic studies of a few congenital diseases demonstrate the ways in which they are layered on our ancient genes, no one has followed up on Taussig's work to study the evolution of congenital heart diseases.[9] We do not yet know whether the ventricular-septal defects in birds are due to similar mechanisms as those in mammals.

We do not even know any more than she did about which congenital deformities birds suffer from. Taussig's catalog of the broken hearts of birds still remains the most comprehensive.

Understanding the evolution of our hearts has allowed us to better understand their deformities and disease. But Taussig did not anticipate how broad the lessons from evolution might be. Considering the evolution of our hearts sheds light on their weakest parts, their coronary arteries.

Today, in most mammals, the coronary arteries are two short arteries that each branch into smaller arteries on which an individual's life depends. If any of them clog, death of the regions of heart it feeds ensues. Often, this is enough to stop the heart; the afflicted suffers chest pain, shortness of breath, and then loss of oxygen to the brain. But even if it doesn't stop, the heart is slow to recover, and the affected muscle is replaced by scar tissue.

These two main arteries, left and right, arise in the aorta and supply blood to the heart. They are the first to branch off the aorta, even before the artery leading to the brain. They are at the root of the phrase *to have a coronary*. But what has long been considered unusual about these arteries is why there are just two to start with, with no backup plan. This, like the congenital deformities, might have an evolutionary explanation as well.

Our human coronary arteries evolved in concert with the evolution of the four-chambered heart. Coronary arteries have evolved again and again in vertebrates to deal with increased levels of activity. Fast, long-swimming fish developed long coronary arteries that run from the gills to the heart (in essence, an extra cycle), for example. But the most conspicuous coronary arteries are found in mammals and birds. Both mammals and birds evolved greatly expanded coronary arteries as the activity and efficiency of their hearts increased. The expanded coronary arteries of birds and mammals were required to deal with the extra activity of the heart in general, but the most acute challenge may have been the left ventricle, which, once it became totally separated from the right, no longer received the slosh of oxygenated blood[10] (though once coronary arteries were present on the right, they were useful on both sides).

Coronary arteries existed in the ancestral lungfish and also in amphibians, snakes, lizards, and turtles, but they were modest in size and flow. They derived from two narrow branches off the aorta, but that was enough. In mammals and birds, the size of the ancestral coronary arteries increased, as did the web of their arterioles and capillaries, but their number, just two, stayed the same. It was easier for evolution to expand these arteries than to make more of them.

In mammals and birds, coronary arteries are now necessary for everything. Our active, warm-blooded existence depends on them. With our hard-charging hearts (and lives), if these arteries get clogged, the heart will not have enough oxygen to pump. It dies. If engineers were designing a heart from scratch, they would give it more coronary arteries for backup. They would arrange things differently. But evolution does not design from scratch; it built our hearts out of lungfish hearts, which were built out of earlier fish hearts, which were built out of sponges' cardiovascular systems. Where mammal species differ at all with regard to coronary arteries, the differences have to do with very small vessels, the collaterals, which run crosswise between coronary arteries. In some species, such as dogs, these vessels are relatively large, whereas in other species, such as pigs, they are nearly absent. In healthy humans, these collaterals carry less than 2 percent of the flow found in the coronary arteries. We depend on the coronary arteries. Our history is our context but also our weakness—we have just two coronary arteries when more would be useful. We are predisposed to die the way we do because our ancestors crawled onto land and evolved warm-bloodedness. Our active success gave us an Achilles artery.

It is this weakness that heart surgeons dealing with clogged coronary arteries confront, a weakness due to our origins, a weakness with its roots in the transition out of the sea. But it gets more complex. The coronary arteries are a weakness only because they clog due to atherosclerosis. In theory, evolution might also shed some light on the heart's weaknesses by helping us to understand when our hearts began to clog. We know that atherosclerosis is as ancient as Queen Meryet-Amun's reign in Egypt. But it could conceivably go all the way back to the first mammal (or bird). Amazingly, this is a possibility no one even considered until a few years ago, when Dr. Nissi Varki and, later, her husband, Ajit, began to look at chimpanzee hearts.

16

Sugarcoating Heart Disease

In 2005 Nissi Varki became deeply intrigued by reports at a meeting of primatologists in La Jolla, California, not far from her house. This intrigue would lead her to a discovery that has changed the understanding of heart disease in humans. At the meeting, scientists from five primate centers, including the Yerkes National Primate Research Center at Emory University, summarized their observations on the causes of deaths among captive chimpanzees. It was, or should have been, really boring, nothing more than a tallying of the final moments of dozens of captive animals.

In the wild, predators, snakes, infections, and other chimpanzees kill chimpanzees. In zoos and research facilities, chimpanzees are removed from most of these threats, and so, it was presumed, they lived long enough to suffer terrible chronic diseases. Yet, although the fates of zoo and laboratory chimpanzees were studied in autopsies, they were not studied from any sort of broad perspective. The literature, to the extent that it weighed in at all, seemed to assume that chimpanzees died of exactly the same diseases in captivity that humans die of in modernity: heart disease, strokes, and cancer.[1]

At the Yerkes laboratory and in other primate centers, predators are kept at bay, diseases are controlled, and the animals eat processed diets (in most cases, Purina monkey chow. Yes, Purina sells monkey chow) supplemented with vegetables and bread. Well-fed chimpanzees move inside their cages, banging around and killing time while exercising much less than they would in the wild. Based on their diet and lifestyles, one might expect captive chimpanzees to have heart attacks, at least occasionally. It did not, then, come as a surprise when talks at the meeting in La Jolla revealed that heart disease was a very common cause of death in the Yerkes chimpanzees—perhaps the most common cause of death, particularly among males, just as in humans. Other studies showed that these same animals also had very high levels of cholesterol. This is not good news if you are a chimp, but it made us, as humans, seem less alone in our plight. When living like us, chimps, our closest relatives, die of heart attacks like us.

Superficially, the hearts of different primates tend to be relatively similar. A gorilla heart looks like the heart of a chimpanzee, and a chimpanzee's heart is similar enough to a human's that in 1964, James Hardy at the University of Mississippi Medical Center successfully transplanted the heart of a chimpanzee into the body of a human patient, Boyd Rush (and Richard Lower later clandestinely transplanted human hearts into baboons).[2] The reality that human and chimpanzee (or baboon) hearts can replace each other, even if only very temporarily, could be interpreted as meaning that our shared ancestors also had hearts like us, including the predisposition to heart disease, given the right circumstances. Heart disease, then, in this retelling, is a potential fate far older than ancient Egypt; it is at least as old as apes, and may extend far back into our mammalian past, to the origin of coronary arteries. An alternative scenario—that chimpanzees and humans independently evolved a propensity for the disease—is less parsimonious (that is, it requires more steps). But sometimes two things that look the same are really different. The meeting in La Jolla energized Nissi Varki to think more about the mysteries of the differences between the diseases of humans and chimpanzees. She decided to temporarily abandon the studies of cancer (in mice) that had occupied most of her career to focus on the pathology of chimpanzees. It was not her first foray into the study of chimpanzees; she had helped with several chimpanzee autopsies in the past. But this was different; this would turn out to be a mystery like none she had considered before.

Often it is said that chimpanzees and humans share 98.5 percent of their genetic code, their DNA sequences, which is true, but a great deal of difference can be found in that 1.5 percent. We know about many of the differences between chimps and humans, differences that evolved rapidly. In the few million years that separate us from the chimps, we lost our fur. We stood upright. Our brains became bulbous, heavy with consciousness. Our feet flattened. Our sweat glands became larger and denser. But the internal features of our bodies—skeletons excepted—are thought to have gone through this historic transition relatively unchanged. They were, the logic was, too fundamental for evolution to tweak; a kidney is a kidney, a liver a liver, a heart a heart—hence the feasibility of cross-species transplants.

But when Nissi started studying chimpanzee hearts, she immediately noticed differences, differences the veterinarians seemed to know about but that had gone largely unmentioned in the human medical research literature. To Nissi, it was clear that at least some of the chimp heart attacks were fundamentally different from human heart attacks. Some of the chimps had suffered interstitial myocardial fibrosis, where *interstitial* refers to the location of the problem (in the gaps between muscles), *myocardial* means "heart muscle," and *fibrosis* is the formation of excess connective tissue.[3] Put it all together and you have the formation of excess connective tissue between heart muscles—the heart becomes bound by its own nonfunctional fibers. While it is not totally clear how myocardial fibrosis starts or kills, one hypothesis is that the disease is triggered by an infection that leads to scarring and fibrosis in the heart and, ultimately, to fatal arrhythmias, in which the heart's beat is no longer synchronized. The fibrosis of the heart muscles prevents the contractions of the heart from moving smoothly from one side to the next, much in the way that oil on the surface of the sea can dampen a wave. Heart attacks due to fibrosis tend to cause sudden death. One moment a chimpanzee is excitedly running around a cage, swinging her arms in the air; the next, she is dead. That such heart attacks occurred in chimps was clear; less clear was how common they were and how they were different from what occurred in the hearts of humans.

Nissi Varki decided to work with veterinary pathologists to study the deaths of chimpanzees at Yerkes along with those at another facility, the Primate Foundation of Arizona,[4] in more detail. She looked at preserved specimens of the hearts of chimpanzees that had died in captivity. She was in luck. Biologists, including those at primate centers, are natural hoarders. They collect everything in the hope that something might someday be useful.[5] The refrigerators and drawers of biology buildings tend to be filled with a miscellany of dead nature: frozen bats, half a woodpecker, tissue samples, a pinecone.[6]

In storage, Varki found samples of the hearts of fifty-two chimpanzees, each one stored in paraffin wax. But before looking at these hearts, Varki repeated what had been done before: she looked at the data the chimp caretakers themselves had recorded as to the causes of the chimpanzees' deaths. In most cases, autopsies had been done, and most of the chimpanzee deaths from 1961 to 1991 were due to infections. But after 1991, once treatment of infections had improved, the most common cause of death, 36 percent of the total (and twenty-one individual deaths), was heart disease.

This was not surprising; it was roughly what had already been reported, with a few more deaths because of the inclusion of slightly more data. But then she and her colleagues examined tissue samples from those diseased hearts, performing what equates to *CSI: Chimpanzee*. The paraffin samples were pulled out of their paraffin, rehydrated, and then stained so as to make a variety of features of the hearts more visible. When this was done, none showed evidence of severe atherosclerosis. The hearts were largely free of plaques, even though cholesterol levels in chimpanzees matched or exceeded those regarded as healthy for humans.[7] Even baby chimpanzees, Nissi Varki learned, have high cholesterol—high enough that if they were humans, they would be prescribed statins. Yet even at these high levels, cholesterol did not appear to be causing most of the heart blockages in chimpanzees. But there was something else. The chimpanzees with heart disease *all* showed evidence of myocardial fibrosis. Fibrosis was even seen in some of the chimpanzees that had died of other causes. Here was a major discovery, one that changed our perspective on the problems of our hearts.

The simplest explanation for why myocardial fibrosis was so common in chimps but so rarely noted in humans was that it had simply been ignored in humans. Maybe we had just been so focused on atherosclerosis that other problems had been overlooked. To test for such a possibility, Varki paired each ape heart sample with a human heart sample. None of the human hearts showed evidence of fibrosis. Chimpanzees and humans both suffer heart disease, but it is not the same disease. So who is strange, humans or chimpanzees? Nissi Varki considered other apes— gorillas and orangutans. The data were sparse, yet nearly every case—two orangutans at the National Zoo, a series of sudden deaths in gorillas—seemed to suggest fibrosis rather than atherosclerosis as the cause. The same for monkeys, where data existed.

What about the closest relatives of primates, rodents? Rodents do not appear to suffer from either form of heart disease, neither myocardial fibrosis nor blockage-induced heart attacks. In fact, even mice with extraordinarily high cholesterol levels, levels that would be regarded as acutely dangerous in humans, do not suffer heart disease (the exception being specialized mice bred to be predisposed to heart disease). It was humans, Nissi Varki concluded,

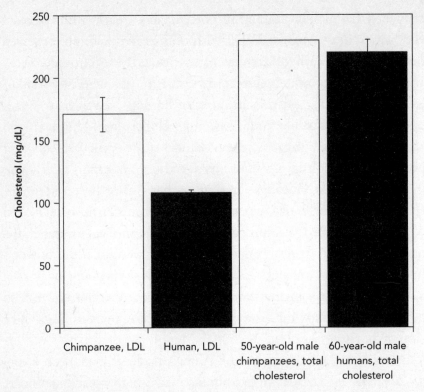

Cholesterol levels in captive chimpanzees and modern humans (living in the United States). The average chimpanzee has cholesterol levels both in terms of total cholesterol and LDL akin to those found in aging humans in the United States. (Data from *Evolutionary Applications* ISSN 1752-4571)

who were strange. As she put it, "Rather than representing a similarity, heart disease is an instance where there are unexplained human-specific differences" from the other apes, differences even from other mammals. We die in an unusual way.

If we are to reconstruct what happened that led our hearts to suffer fates different from those of living apes, we need to return to the evolutionary tree. Among living primates, humans are most closely related to chimpanzees and bonobos, their libidinous cousins, from whom our ancestors diverged about five million years ago. The branch that includes humans, bonobos, and chimps branched from that containing gorillas about eight million years ago (which branched from that with orangutans and gibbons even more remotely, twelve and fifteen million years ago, respectively). If all the other apes suffer from myocardial fibrosis while humans alone suffer from atherosclerosis, there are two possible explanations. One is that each of those ape species (and the monkeys and the mice) independently evolved some feature of the heart or immune system that led to fibrosis. The other is that human hearts have evolved special attributes that, while they reduce (or eliminate) the risk of myocardial fibrosis, predispose us to an equally fatal fate. Of these two options, parsimony suggests it is far more likely that the latter is true—that our species is the odd one. If I were to write a book about chimpanzee hearts or gorilla hearts or mammal hearts more generally, I wouldn't even need to mention atherosclerosis or clogged hearts except to note how strange humans are.

Nissi Varki needed to account for not only the absence of the ape form of heart disease in humans but also the presence of the uniquely human form of heart disease we suffer. She can't explain the heart disease of chimpanzees yet. No one can, nor has anyone really tried. If you would like a mystery to solve, the heart disease of chimpanzees and other apes awaits you (I would start by looking for pathogens in the chimpanzee hearts if I were you). But we can't ignore our own fate. It is clear our heart disease is not due to having too much cholesterol. Instead, it appears to be due to the body's response to that cholesterol. Cholesterol flows freely in chimpanzee blood without forming plaques. The plaques in our blood and hearts are due to the immune system's response to cholesterol and other substances. The human immune system reacts to cholesterol as though it were foreign and smothers it with immune cells called macrophages. Plaques are cholesterol in LDL buried in the artery walls by the accumulation of macrophages. We need, then, to explain why our bodies attack cholesterol while the bodies of the

rest of the primates do not. We need to understand the recent evolution of our immune systems to understand our hearts.

The answer to Nissi Varki's mystery came from her husband, Ajit
Varki, with whom in recent years she had begun to collaborate. Ajit
and Nissi Varki met at the Christian Medical College in Vellore,
the preeminent medical college in India. There, they fell in love
and, simultaneously, began to specialize professionally. Ajit was
interested in internal medicine, hematology and oncology in particular, and Nissi in pathology. Nissi had already begun to study
mice as models of humans to understand the biology of cancer and
other diseases. Ajit researched practical treatments for common
diseases. Neither of them studied the heart, nor did they imagine
they would study the heart in the future. Ajit finished his degree
before Nissi and went to the United States in search of new research
skills. He took a job first at the University of Nebraska and then at
Washington University in St. Louis. Nissi finished her degree and
joined Ajit at Washington University, where she too found a
position; eventually, they both moved to their current jobs at the
University of California, San Diego.

Early in his career Ajit Varki found himself working on compounds called sialic acids. He started in 1984, accidentally. Ajit was
treating a patient with a relatively rare blood disorder, aplastic anemia, by giving him a derivative of horse serum.[8] But there was a
problem: the patient's immune system reacted to the serum, causing a rare serum sickness. Ajit determined that this particular
serum sickness was due to a reaction of the human body to one of
the forms of sialic acid in the horse serum.[9] Sialic acids are sugars
that coat the surfaces of cells, sometimes as densely as hundreds of
millions per cell. The patient's immune system was responding to
the sialic acids in the horse serum as though they were dangerous,
foreign entities. This was a very unusual reaction, "ridiculous"
even, as Ajit would later say, because all mammals, all vertebrates in

fact, were known to have the same exact forms of sialic acid on their blood cells, the two most common of which were Neu5Ac and Neu5Gc (note that the difference is just an *A* relative to a *G*, but much depends on this one-letter difference).[10] Humans and horses should both have these same two sialic acids. The immune system of a human shouldn't even have noticed that the horse serum was different. Some part of what was known about sialic acids, horses, and humans was wrong, but it wasn't obvious just what. It was, as Ajit told me in an interview, "a kind of detective story."

A decade passed, during which Ajit immersed himself in studying sialic acids and other sugars on the surfaces of cells; he became an internationally known expert on the topic (glycobiology), literally writing the book on these sugars.[11] He and Nissi collaborated for the first time, briefly, to understand sialic acids in mice (lab animals could be studied much more easily than either horses or humans).[12] Even though it was never the entirety of what Ajit worked on, the horse-and-human mystery stuck with him. He began to accumulate clues. The first clue came when he found that earlier studies had documented that humans seem to lack one of the two main kinds of sialic acid, Neu5Gc, found in other mammals. On its own, this was unusual. Then he found a study published in 1965 showing that some nonhuman apes had this normal sialic acid (Neu5Gc), an observation soon confirmed by Elaine Muchmore, also at the University of California, San Diego, with whom Ajit had begun to collaborate.

On the basis of these clues, Ajit Varki and Muchmore decided to compare the genes associated with the production of sialic acid in sixty humans and all of the apes. While it was known at the time that about 1.5 percent of the sequences of genes in humans and chimpanzees were different, none of those differences had yet been pinpointed. Varki and Muchmore would be the first to do it, and they were in for a surprise. In all of the apes (and the rest of the other mammals so far studied, including mice and horses), an enzyme modifies the basic sialic acid, Neu5Ac, by adding an oxygen atom to it, which converts it to Neu5Gc (*N*-glycolyl-neuraminic acid). The gene that produces this enzyme, CMAH, is broken in humans,[13] missing ninety-two bits (nucleotides) of DNA. As a result, all sialic acid produced by humans is the Neu5Ac form and lacks the extra oxygen atom. Humans lost the ability to make Neu5Gc. Because sialic-acid sugars are found on all of the cells in the human body, this was enough of a difference to cause the human immune system to see the Neu5Gc in the horse serum as different.[14] Every cell in a human is different from every cell in every other mammal. Humans are the odd species out.

Ajit and his colleagues had discovered the first genetic difference ever noted between humans and chimpanzees,[15] a difference that, as he learned from the very beginning, made human immune systems different from those of chimpanzees but also nearly all other mammals, including the horse. We fixate on the conspicuous differences between chimps and us, but in a world filled with pathogens and the diseases they cause, the inconspicuous differences may be far more important. Sialic acid, Ajit was nearly sure, was an important piece of the story of human evolution and disease.

Ajit Varki wanted to pursue this issue further, but he needed to learn more about chimpanzees and other apes. The difference between chimpanzees and humans seemed fundamentally important, but it was unclear why it existed. He needed to know more about chimpanzees to make sense of the observation but also to understand what other differences he and everyone else might be missing.

Ajit took a sabbatical to spend time at the Yerkes primate research center, where Nissi would later make her observations about chimpanzee hearts. While there, Ajit did not focus on the heart. But he was able to document a list of diseases that seemed to differ between chimpanzees and humans, diseases that might somehow be associated with the change in sialic acid and recently evolved differences. The list included some cancers, AIDS, a range of infections, rheumatoid arthritis, and much more. And so, years later, when Nissi came home from Yerkes with her revelation about the differences between human and chimpanzee hearts, it was not a total surprise to either of them (though it would be to nearly everyone else). In thinking about the heart, Nissi and Ajit started to talk about sialic acids and Ajit's earlier trip to Yerkes. It is ridiculous to imagine that the sugars that Ajit had spent decades studying would prove to be the answer to Nissi's mystery about chimpanzee hearts, and yet that is precisely what happened.

In order to put these two stories together, what the Varkis needed was one final clue, and fortunately, it was one Ajit Varki had already turned up. When present in the human body, the normal mammalian sialic acid (the one with the extra oxygen) causes an immune response; the human body notices the extra oxygen and attacks. But something was odd. From what Ajit knew about sialic acids, it seemed that when humans consumed mammal meat, the sugars on the meat would get incorporated into human cells. If this was true, it would make some human cells, those with the sialic acid with an extra oxygen, look to the immune system as though they were foreign. The immune system might overreact and attack these cells, leading to all sorts of problems, including, and here was Ajit's speculative jump, atherosclerosis. But Ajit did not have any proof of the theory that the sialic acids in human diets were getting incorporated into human cells, and so he wanted to do an experiment.

Ajit Varki wanted to feed humans mammal meat and see if the sialic acids from that meat (Neu5Gc) ended up on the cells of the humans. He couldn't do the experiment on any other mammal because humans are the only ones who lack this particular mammalian sialic acid. He thought it would be easy enough to do on himself. But universities are very reluctant to allow scientists to do self-experiments (times have changed since the days of Forssmann). So before even proposing an experiment on humans, Ajit did one on human cells grown in petri dishes in the lab. When fed the normal mammal sialic acid (Neu5Gc), those cells incorporated it directly into their membranes! After seeing these data, the university review board allowed Varki to do an experiment on himself. Varki and one of his collaborators, Pascal Gagneux, then extracted sialic acid from pig salivary glands (sialic acids are very concentrated in spit). On the morning of February 16, 2001, Varki checked into a clinical research center at his university. Once there, Varki consumed an amount of sialic acid equivalent to eating "fourteen pork steaks," 150 milligrams, in the form of a pig-spit Slurpee.

Over the next weeks, the levels of Neu5Gc sialic acid in Varki's urine, saliva, and hair increased. This mammalian sialic acid was becoming part of his cells. The experiment was later repeated on Gagneux and Muchmore, with similar results.[16] Like the old adage says, Varki became what he ate.

People who eat mammal meat incorporate the sugars from the meat into their cells. The body's immune system sees the sialic acids on the tips of these sugars as foreign, and it attacks, a reaction that occurs throughout the body, including in the artery walls. The fact that this occurs is unassailable. This, for the Varkis, was the final clue related to Nissi Varki's discovery of the differences between chimpanzee and human hearts.

Over dinners, lunches, and more formal meetings, the Varkis concluded that the differences in human sialic acids compared to those of other mammals combined with a diet rich in mammal meat contributed to the prevalence of atherosclerosis in humans. What is unclear is how much of human atherosclerosis is due to the loss of one type of sialic acid. The human immune system is unusual in its activity level, which is high, even without the influence of sialic acid from other mammals (vegetarians are not immune to atherosclerosis). Relative to other primates', the human immune system is twitchy and overreactive.[17] Together, these two factors—the lack of a particular sialic acid and a general twitchiness—may be at the heart of many diseases. It is this overreactivity that may cause humans to develop AIDS when infected with HIV (chimps contract HIV but do not develop the full-blown disease). It is this overreactivity that is associated with chronic hepatitis B and C, rheumatoid arthritis, asthma, type 1 diabetes, and other modern plagues. In other words, all humans appear to suffer more inflammation (in general, but particularly in the arteries) than chimpanzees do, but those who consume mammal meat may suffer it disproportionately.

All of this paints a compelling picture linking sialic acid to human immune response and diets. But it doesn't answer one question: Why are we unusual in our sialic acid and immune reactivity in the first place? For this, too, the Varkis have an idea. Humanity's most recent common ancestor, from whom all humans inherited at least some genes, lived at least a hundred thousand years ago. Any problems we all share, any uniquely human traits, are at least that old, and so if we are to discover what makes human hearts unique, we must go back that far. From the very beginning, it appears our ancestors were different from their close relatives in terms of something besides sialic acid, before the change in the sialic acid: our ancestors were plagued by disease.

Earlier work on which Nissi and Ajit collaborated showed that sialic acids played a key role in human interaction with many of the pathogens that cause disease. Pathogens such as the influenza virus use sialic acids to identify particular cells in the upper respiratory tract and latch onto them. Perhaps, the Varkis began to think, the change in sialic acid was an attempt by the cells within the bodies of our recent ancestors to escape some particular infectious disease.

The story of humans and disease is unlike that of any other mammal and disease. At some hard-to-define point very near to the origin of our kind, humans combined language skills, brainpower, and the beginnings of culture in such a way that allowed larger groups to live together than had ever lived together before. The largest known nonhuman primate group ever recorded contained about fifty individuals (evolutionary biologist Mark Moffett has even called this the fifty rule). Scholars aggressively debate exactly when this transition happened, why it happened, and how it changed humans, but what no one debates is that, as humans gathered in ever larger numbers, the risk posed by old and new pathogens increased. We are familiar with this phenomenon from schools. Put all the kids together, and pathogens and parasites—whether it's the flu, norovirus, lice, mites,[18] or even one of the viruses causing pinkeye—spread rapidly. Put our ancestors together, and the same thing happened. Most primates host about two hundred species of pathogens; humans now host more than two thousand species, two thousand different and potentially dangerous and deadly forms.

The greater density of human populations meant that pathogens could spread faster; it also meant that they could become more deadly. Pathogens do not generally evolve to be very deadly to their hosts because if they are too deadly, their host dies before they can get to another host, but there are a number of dissatisfying exceptions to this satisfying generality. One has to do with the density of the hosts. Once humans began to gather, the chances of a pathogen getting to another host improved and so the costs of deadliness declined. Perhaps some widespread and virulent pathogen, in these early moments of our human story, triggered evolutionary changes like those the Varkis found, changes in sugars throughout the body. If such a pathogen evolved, it might have killed so many of our ancestors that only those with versions of genes that allowed them to escape those pathogens would have survived. Maybe one of those protective genes was a broken version of the gene for Neu5Gc sialic acid.

Prior to the past few hundred years, our ancestors died of many things, but among the most common were diseases caused by a suite of species that lurked in the blood. It is these species that the Varkis think altered human bodies in ways that ultimately made heart disease more likely.

The idea that those pathogens that make their way to the blood might alter the evolution of the body is not far-fetched. Blood is the stuff that is simultaneously most precious to our bodies and most attractive to other species. Blood is the heart's liquid, the ether through which the heart asserts its influence over the body. Blood is precious to our bodies' cells. It has water where water is scarce, proteins, sugars, a reasonable pH, and a constant temperature. But the preciousness of the blood is also its weakness. Any parasite that gets access to the blood finds its own piece of an immense liquid paradise. Each adult human body holds about a gallon of blood. With about seven billion people on Earth, that is about seven billion gallons of blood. No wonder, then, that no fewer than twenty lineages of flies have independently evolved a fondness for blood, accompanied by massive changes in parts of their mouths that allow them to pierce skin, forcing blood to flow freely so they can do the job quickly enough to get away. Some bats feed on blood, as do leeches. One vampire finch even feeds on blood. A finch! Recently, a moth with a fondness for blood was found landing wantonly on salty arms in Siberia. But whereas these animals stopped at the skin—braced their feet and bit in—others, including the pathogens that began to latch onto human society as our groups grew in size, figured out ways to press their entire bodies through the flesh and into the stream of pulsing sustenance.

Nearly all of the species that make it into our blood influence both our blood and, via the blood, our hearts. The species whose impact we best understand is the malaria parasite, *Plasmodium falciparum*. Malaria kills more than a million people a year today, and that is many fewer than it killed before pesticides, mosquito nets, and prophylactic medication. A malaria parasite rides in a mosquito until the mosquito lands and punctures skin. When the mosquito jabs in, the parasite slides with the gush of saliva into the blood, where it rides to the liver, divides, and produces more parasites; these break free, travel back to the blood, and invade red blood cells. The body reacts with fever, attempting to kill the parasites with heat. But by the time the parasites have made it out of the liver and into the red blood cells in which they ride, again and again, through the heart, it is too late. When another mosquito comes along, one of the parasites, having helped consume its host into near disaster, catches a ride to the next body. In this way, malaria has moved from person to person around the world. In many human populations ten thousand years ago (even two hundred years ago), most people contracted malaria between birth and death. Perhaps one in ten of them died from it.

Chimpanzees and gorillas, like us, suffer from malaria. Nearly half of all chimps and gorillas are infected with malaria parasites at any given moment (bonobos, interestingly, appear to harbor none; why that is the case has not yet been studied). But their malaria is different from ours, seemingly less deadly. Humans are known to have contracted one form of malaria by way of a gorilla about twelve thousand years ago, with the dawn of agriculture. A mosquito bit a gorilla with malaria; a malaria parasite from the gorilla hitched a ride on the mosquito, which then bit a human. The parasite arrived in the human and reproduced, and the species began to evolve in ways that allowed it to take better advantage of its new host. (The malaria parasite is not the only thing we have acquired from gorillas. Genital lice also appear to have made the jump, though that would have had to happen when a gorilla ancestor and a human ancestor were—ahem—touching.) Such host shifts are common, but the strange part of this story is why, prior to twelve thousand years ago, humans did not already have their own unique malaria parasite, why we ended up with a transmogrified and very deadly gorilla malaria. Chimps had chimp malaria parasites, gorillas had gorilla malaria parasites. Where was the human malaria parasite, the parasite that evolved to live in us before we picked up gorilla malaria? The Varkis posit that an ancient human malaria was missing because our ancestors had evolved a way to escape it.

Perhaps that happened once human settlements became more dense and human malaria more deadly, and perhaps it happened because human malaria parasites (like chimp and gorilla malaria parasites) latched onto Neu5Gc, which was the form of sialic acid that our ancestors lost. If an absent or broken Neu5Gc made some of our ancestors immune to an ancient human malaria, the genes for this trait would have swept through human populations, potentially causing this parasite to go extinct. If that did occur, the reprieve in human malaria was, unfortunately, temporary, lasting only as long as it took our kind to be colonized by gorilla malaria.

This idea is plausible. We know that malaria can shape our genes; the colonization of humans twelve thousand or so years ago by the new malaria strain (which, in shifting to humans from gorillas, also had to shift the sialic acid to which it bound) clearly has. One of the changes that malaria caused in our genes was to favor versions of human genes that made it harder for malaria parasites to be able to persist and/or breed in our blood. These changes are primarily deformities of the hemoglobin in red blood cells. The deformities make the red blood cells less effective at holding on to oxygen and so tend to cause the heart to become more muscular and beat faster. These changes are actually negative changes, problems, but they are smaller problems than dying of malaria. It has been argued, quite plausibly, that the O blood type evolved as a response to malaria. Other mammals have A and B blood types, but not O. Individuals with O blood types lack certain sugars on their cells, sugars that make it easier for malaria parasites to find and enter the cells. As a result, individuals with the blood type O have a lower risk of dying of malaria. All of this together points to the possibility that malaria shaped human evolution in other ways as well, as any of the deadly pathogens of our history might have.

In this telling, the loss of the primate version of sialic acid is an ancient adaptation that is no longer useful now that we have been colonized by gorilla malaria (or, in a few lucky places on Earth, escaped malaria altogether). It may be that evolving a broken sialic acid gene allowed humans to escape a number of pathogens at once, at least for a while.[19] But this escape left us with overreactive immune systems, which, when combined with poorly designed coronary arteries, diets rich in mammal meat, sedentary lifestyles, smoking, and other risk factors, began to clog arteries and kill us. The combination of these historical weaknesses, though, did not begin to kill humans with great frequency until relatively recently, for the simple reason that humans didn't live long enough.

Modern apes live just fifteen to thirty years in the wild.[20] Six million years ago, when our ancestors diverged from the ancestors of chimps, they probably lived no more than thirty short and brutish years, punctuated by episodes of eating fruit, fleeing from leopards, mating, and then, postmating, savoring some more fruit. Every so often, an individual might live much longer, but rarely. Inferring what happened between six million years ago and today is the difficult part.

Our best guess as to the life expectancy of early humans comes from studies of hunter-gatherers alive today, who tend to live around forty years (on average — such reconstructions always focus on averages), and we can infer that at some point in the transition from chimplike ape to hunter-gatherer human, our species gained some twenty or so years. Human life expectancy hovered around forty years through most of the nine hundred thousand years of hunter-gatherdom. If there is a natural human longevity, this is it: about forty. Then, as humans began to farm, life expectancies appear to have changed again; they seem to have decreased slightly.[21] The royal mummies in the Egyptian Horus study, for example, had an average age of just thirty-eight years. No telling how long the average Joe lived, though everything we know about the long history of haves and have-nots suggests it is likely to have been fewer years. It was not until the 1800s that human life expectancies began to creep up into the fifties and then sixties and then seventies, and even, in recent decades, the eighties, at least in developed countries. In much of the world, a sixty-year-old is still an ancient.

Focusing on averages skews our understanding of these transitions to some extent. Averages include infant mortality, and one of the biggest shifts through time has been a progressive decrease in infant mortality (which removes many of the data points where life expectancy is less than one year from the averages). But even once infant deaths are taken out of the picture, human longevity generally increased through time. This means that, as time went on, more individuals were at risk of heart disease, even if nothing else about their lifestyles or biology changed. Today, heart disease and strokes (as well as cancers) tend to kill humans only when they are older than thirty-five. Until two hundred years ago, humans were usually killed by disease, predator, or mishap before their hearts and blood vessels had a chance to fail. The Egyptians, for example, do appear to have suffered greatly from heart disease, but very few of them—Queen Meryet-Amun being an interesting potential exception—lived long enough for heart disease to kill them. Human bodies have not had a chance to evolve a response to the overreactivity of their immune systems, and the more successful the medical treatments of immune disorders (including heart disease) are, the less likely it is that they ever will.

The Varkis continue to try to understand this story; many more details will probably emerge, but the generality seems likely to remain: we have atherosclerosis in part because our immune systems react to the LDL in our blood. Our immune systems are more aggressive than those of other primates because we were (and, in much of the world, still are) plagued by an unusually high diversity of pathogens, including pathogens deadly enough to cause an alteration in our genes. As a result, any lifestyle that provides more cholesterol in LDL in the blood also provides more substrate for the immune system to attack. By contrast, any lifestyle that increases HDL will decrease LDL, reducing what can be attacked.

The consumption of foods rich in antioxidants reduces the proportion of LDL molecules that are damaged by age, oxidized, and then attacked; anything that heightens immune reactivity makes attacks more eager and problematic. Smoking, as I've mentioned, triggers inflammation and an immune response, and it constricts arteries. There are a diversity of ways in which inflammation, cholesterol levels, the relative abundance of different types of lipoproteins, and the oxidation of lipoproteins influence the risk of atherosclerosis, and there are probably many more contributing elements that we don't yet know about. Already, there are more factors involved than Ancel Keys or Akira Endo could have possibly imagined, factors influenced by our modern behavior as well as our ancient evolutionary history and our relationships with other species.

Take, for example, the ecology of teeth. Modern mouths are susceptible to tooth decay and gingivitis due to oral pathogens such as *Streptococcus mutans* and *Porphyromonas gingivalis*. But this, it turns out, is a relatively new state of affairs. Studies of the plaque on the teeth of ancient mummies — one of the few places where the DNA of ancient human-associated bacterial species can be determined with any certainty — have revealed at least two major transitions in oral ecology, both related to carbohydrate consumption. First, when humans switched from hunter-gatherer diets to the mealy foods of agricultural diets, gingivitis bacteria in their mouths increased. This increase led to more inflammation, more gingivitis. Second, over the past two hundred years, industrialization of the production of sugar, and its subsequent increase in human diets, has led to a reduction in the number of kinds of bacteria in the mouth. Sugar has favored the dominance of a few bad-news forms of oral bacteria, including *Streptococcus mutans* and its kin, which cause dental caries. These changes might not seem to have anything to do with the heart, but they do. Chronic gingivitis and tooth decay make the immune system more active, leading it to produce more and more macrophages, which must travel through the blood. In traveling through the blood, these macrophages meet up with the cholesterol in LDL, which they attack. In other words, the changes in human diets over the past twelve thousand years affect the cardiovascular system even when those changes involve areas far removed from the heart.[22]

Eating too many simple carbohydrates increases heart disease by increasing oral microbes (as well as triglycerides). Meat-eating contributes to atherosclerosis in more ways than just those noted by the Varkis. As Keys observed, eating meat high in saturated fats affects the levels of LDL and HDL in the blood, though the degree to which this happens is genetically dependent and the subject of

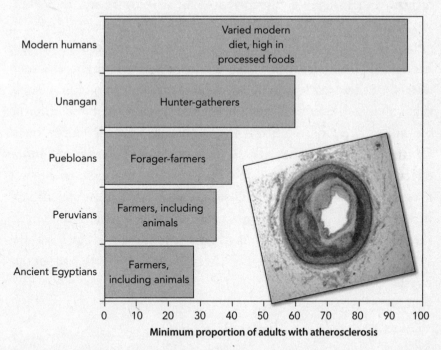

The proportion of individuals with evidence of atherosclerosis in at least one artery. Estimates for mummies are minimum estimates because not all arteries are visible in most mummies. Data are from *The Lancet* 381, no. 9873 (April 6–12, 2013): 1211–12.

great debate. But meat-eating also increases inflammation, because the sialic acid in nonhuman animal meat triggers the response of the immune system (perversely, this means that the one kind of meat that can be eaten without incurring the chronic wrath of one's immune system is human meat. Cannibals might be expected to be at a slightly reduced risk of atherosclerosis). Finally, a recent study has shown that individuals who eat red meat tend to have different bacteria in their guts; these may be the kinds of bacteria that cause problems. Red meat and eggs both contain the compound choline (as well as other compounds related to choline). Choline is necessary for all animals (and its absence in the diet can cause heart problems). However, some bacteria found in the guts of meat-eaters turn choline into another compound, trimethylamine, the presence of which increases the formation of atherosclerotic plaques.

The effects of all these factors depend on genes associated with cholesterol, inflammation, and even a predisposition to certain bacteria. This makes it clear that while we humans might be able to decrease atherosclerosis and heart disease, doing so is complicated. Our modern heart problems are not simply the result of eating the wrong foods but of living long lives in bodies built for shorter ones. Our bodies are complex, filled with thousands of species and influenced by many more. Our bodies have evolved to escape pathogens long enough to make children, no matter what that escape costs us in old age or in other contexts. Our bodies have life expectancies shaped by the realities of human history, not our hopes for the future. Nowhere, it turns out, is this clearer than in the rate of our hearts.

17

Escaping the Laws of Nature

We repair ourselves — we parry every blow, and all in the service of keeping the gyroscope of life spinning smoothly.

— SHERWIN B. NULAND, *THE WISDOM OF THE BODY*

In medieval Christianity, God was said to inhabit the heart, taking notes on its inner walls. To some, this writing was literal.[1] The heart was a muscular parchment on which the Almighty scribbled vigorously, recording each furtive glance, each generous or ungenerous thought. A long life would fill the heart with stories by which one's worth could be fairly judged. With time, the heart was seen in sufficient detail for men to realize there were no notes inside. To the extent that our stories are recorded in our bodies, we tend to think the records can be found in our brains. The brain does indeed record our actions to memory, albeit via chemistry. But among one group of scientists, a group that has taken an entirely different perspective on our tickers, the heart is still a record of our fates in this life rather than the next.

You can probably think of many things the heart rate says about us. The heart's pounding and thumping enthusiasm speaks of fear, of love and lust, of how much one exercises, of the lives of one's ancestors, of one's diet. Some scientists argue that it also speaks of our demise — or, rather, a comparison of the human heart to those of other animals does.

These scientists study what is called scaling (also referred to, even more obscurely, as allometry), a field concerned with understanding the relationships of different features of organisms as they grow or evolve. Scaling tells us how big dinosaurs would have been able to get without collapsing under their own weight and why plant-eaters tend to be bigger than predators. It also tells us how tall the tallest trees can grow and how densely trees (or cells) can live. Many of the phenomena that scaling purports to explain are straightforward and well understood. Although life is inscrutably diverse and complex, scaling reminds us that such complexities tend to obey the universal demands made by physical laws. Nobody gets away from gravity or momentum.

The laws of scaling are said to follow those of physics, and so perhaps it is appropriate that one of the most ambitious and reaching allometricians, Geoffrey West, spent most of his career as a physicist at the Los Alamos National Laboratory, the research center established to develop the Manhattan Project. West worked on the theoretical physics of protons and neutrons at Los Alamos from 1976 until 1993, when the $11 billion necessary to build the Superconducting Super Collider in Texas, where he had planned to do his future research, failed to fully materialize. West was fifty-three. He could have taken it a little easy and drifted into retirement, but instead, he decided to try something new. He had done what he could with protons and neutrons; he wanted to explain life, including the bustle of human life in cities and the bustle of blood in the heart. He eventually went to work at the Santa Fe Institute, where he was given a big space in which to do his job, which was to think.

Think he did. West is a mathematical thinker, the sort of character who appears in movies but seems unlikely in the real world. Yet he exists. He is a tall, thin, angular man, all elbows and knees, with untended silver-gray hair and a beard and mustache of the same color. He reaches his arms out, hands open, as he talks, and his weedy eyebrows rise as he makes a point or ponders. He thinks. He talks. He writes equations and codes models. The son of a professional gambler and a dressmaker, he tries to weave beautiful ideas by playing the odds. He uses mathematics to see aspects of the functioning of the universe that others have missed. He is drawn to elegantly simple explanations for complex real-world problems, and he uses equations to describe where mere words fail. In doing so, he is always "onto something," the next clear view of some muddied problem. The possibility of the next clear view, like the possibility of climbing the next big hill, is great enough to sublimate many ordinary necessities. West gets by, when he is working, on a diet of nuts and tea. He can't be bothered to deal with more. Maybe, as he told one journalist, he is allergic to food. Just as likely, he is allergic to worrying about mundane things when the next equation or analysis might explain something as central as the function of cities or, better yet, the human heart. As he told that same reporter, he wants "to find the rules that govern everything."[2]

The work West has become best known for is the study, begun in 2003, of how humans live and act in cities. West had the idea that much of the raucous detail in cities, detail that seemed whimsical, the pure art of living, was predictable, in the way that, as a general rule, the behavior of neutrons are. Sure, an individual neutron or person might make wild deviations from the expected, but collectively, the behaviors of humans and neutrons are predictable, he believed. West thought that if he knew the size and demands of a particular city, he could predict its necessary infrastructure as well as its major human features. The people in a city need resources, and these needs are dependent on the number and density of people; they must, in the terminology of the field, scale with them. As a result, West thought, the relationship between the demand for these resources and the size of the population must obey certain rules, since roads can only be so wide, bodies require a certain amount of food, and buildings need a certain amount of energy. But what if there were also rules for how population size influenced not just what humans needed but also what they did, rules that governed, for example, where art and innovation thrived? If so, the study of these relationships might explain apparently diverse and unconnected things across geography and history: how many miles of road there would be in a city of 150,000 people, how many hot-dog vendors, how many robberies and of what type, but also how much artistic and scientific innovation would exist. From simple mathematical rules, an entire circus of particulars can emerge, or at least that is what West hoped. He hoped he could weave all of the beauty of human creation into math.

To test whether the features of cities were predictable, West collected data on cities. He considered anything that might be predictable, anything that might depend on the "how many" and "how dense" of our living ways. He and his team gathered data like magpies collecting shiny pebbles, avidly and somewhat indiscriminately. They would look at any features of cities they could find information on. Whole fields already considered how cities worked—urban planning, architecture, landscape design. West's team was large and included experts in multiple fields, but they would ignore these fields. West's team would build on physics and biological scaling rather than on the work of the thousands of individuals who had spent their lives studying cities.[3] It was an act of arrogance suitable to a man who sometimes compared his own efforts to those of Galileo.

Amazingly, after years of gathering and modeling, it has all worked out to West's great satisfaction. He can, he thinks, explain cities, and his explanation steadily becomes more useful, since each generation is ever more urban and so ever more subject to West's laws. If you tell him the size and population of a city, he will predict for you the surface area of its roads, the number of gas stations, the scope of the sewage system, and then also its human phenomena, such as the pace of innovation. But the interesting thing is not just that he can predict these features, but that each of these features relates to the size of a city in a slightly different way. The resources a city needs (gas, food, and so on) decrease on a per capita basis as the size of a city increases (bigger cities are progressively more efficient; the slope is less than 1), but the innovation a city engenders shows the opposite pattern: it increases on a per capita basis as cities become larger (the slope is greater than 1). In short, the bigger the city, the greater its efficiency and innovation; the more art and the less gas.

One should be skeptical of anyone who attempts to explain modern human life with a few equations, but West thinks he and his collaborators have solved the city. Many argue with the details of West's models and approach. He trades, they say, accuracy for comprehensiveness. West's equations have a sweeping simplicity that is frustrating to those in the detail-rich trenches.[4] His models do an okay job of accounting for many features of the world rather than a great job of predicting or understanding any particular feature. Others suggest the interesting things about West's models are not where they work, but where they fail, where cities violate his laws. But cities are not the only realm West has considered. He actually started with the study of human bodies and human longevity; in some tellings, this is part of why he began to study scaling in the first place. He was thinking about his own death. West comes from a long line of working-class men who die young, and he had begun to consider the possibility that his own death might not be far off. He wondered about his heart, a heart that, in its beating, was not so unlike the throbbing center of Los Angeles or New York. He wanted to know when it would stop, when he would die, when any of us would die, and why. In hearts, West saw physical laws in action. In hearts, he found what he described as "the single most pervasive theme underlying all biological diversity," all life— namely, death.

When West began to consider the general laws of the heart, he did so, as he later would with cities, naively. It was, he said, "like learning about sex on the street."[5] He read up about the heart in a high-school textbook, the only thing lying around. It was enough to get him excited. Then he began to read the older literature on scaling and the body and he became fascinated with the ways in which hearts differ among peoples and species. West found that heart rates varied widely among animals. Yet this variation followed a simple pattern: *the larger the animal, the slower its heart.*

This idea is not new. Any hunter knows the hearts of smaller mammals tend to beat faster than those of larger ones. If you catch a mouse and lift it up, you can feel its tiny heart zooming in your hands. If you ride an elephant or a big horse, its heart beats slowly and deliberately against your legs. What is relatively new is the sophistication of the explanation for this pattern and the awareness of its universality. The explanation comes in nested pieces.

In articles by his intellectual antecedents, such as Max Kleiber, among dozens of others, West read observation after observation about heart rate that appealed to him. Generations of biologists had studied the metabolism of the human body, its energy use, from the perspective of physics. At the most basic level, these biologists noted that the hearts of smaller animals beat faster and each of their individual cells used more energy. Larger animals were like bigger cities: they simply used less energy per capita (where the *capita* was a cell rather than a person).[6] The existence of this relationship had been well explored. It had even been described as a law, the quarter-power scaling law, a reference to the slope of the relationship between body size and metabolism. It is on the basis of this law that drug doses, whether statins, beta blockers, or anything else, are calculated. But its cause remained baffling; many theories had been offered, typically having to do with the relative ease of heating a larger body as opposed to a smaller one, but none of them quite worked. The attempt to explain how and why metabolism, body size, and heart rate related to one another had reached an impasse. West picked it up as if it were an abandoned toy and started playing with it.

West began with the physics of cells and blood vessels in order to work his way back to the heart (and maybe even to the question of how long an animal like him would live, a question that leaped like a hard-to-catch fish through the marginal waters of his mind). Arteries were like highways, capillaries like back alleys, and blood cells like food trucks, he thought. Their features must relate to the size of an organism, just as their parallel features in a city related to its size. This was not new to the world, just to West, and yet it was compelling. He read as much as he could find about the biology of the body. He met biologists and drained their brains of anything that seemed like it might be of use. When this was not enough, he started collaborating with two of these biologists, Jim Brown and Brian Enquist, then both at the University of New Mexico (and both as ambitious and far-reaching as West). Together, the three began to see patterns they thought others had missed, a kind of underlying mathematics of cells, vessels, and pumps. With these patterns in mind, they used their computers to build a simulated system of blood vessels, a fractal system in which each main branch diverged into ever smaller branches until they arrived at the smallest vessels, the capillaries. The scientists wanted to devise the simplest computer model of blood vessels that could account for the differences among types of organisms. As they started to build this model, they stumbled upon an empirical observation that would prove important, a way in which evolved bodies differed from simulated ones. In theory, a model of the blood vessels could produce infinitely smaller and smaller branches so that a (simulated) big animal could have enough capillaries to get to every cell. But what the trio noticed was that actual human capillaries did not get smaller and smaller. In fact, all of the capillaries in a body are essentially the same size. More than that, capillaries in different animals all seem to be that same size too (just as the size of the smallest vessels are invariant among plants). The capillary's size is set by the width of a blood cell (capillaries are all one cell wide). This means that the capillaries of a shrew, for example, are much larger relative to its body than the capillaries of, say, a blue whale.

The sameness of capillaries meant that there were physical limits to blood vessels reaching cells, limits very similar to those present in the streets of cities (which never get narrower than one car wide). As animals get bigger and bigger, their body volumes increase, and the number of capillaries must also increase in relation to that volume. But as it does, something else happens. As the number of capillaries rises, the volume of blood increases, since blood has to get to all of the capillaries. As a result, the heart of a larger animal must pump much more blood. But even though the size of the heart (and its major vessels) can get bigger and bigger, the ability of the heart to get oxygenated blood quickly to every cell in the body decreases. The blood vessels must branch more and more times to meet the demand of the larger volume, and so it takes longer to get the blood throughout the body, and the concentration of oxygen in the farthest capillaries declines. Because it also takes longer for the blood to get back to the heart, the heart rate slows, as does the metabolic rate of each and every cell of the body (which is why larger animals are, all other things being equal, a little cooler than smaller ones). This was, it seemed to West and his new friends, the mechanism lurking behind the ancient relationship between body size and metabolism. Just as in cities, it was a constraint of how roads work; there are relatively few ways of getting from here to there when the width of the path is unalterable.

Based on an understanding of the capillaries and a model of the fractal connections of blood vessels of larger and larger diameters, West, Enquist, and Brown can predict, in a simulated world, not only the metabolic rate of a mammal but also its number of capillaries, aorta size, and heart rate as a function of its size. What is more, the slopes of the relationships broadly match up with those seen in nature. Add a factor to account for body temperature, and they can predict those same features in birds, reptiles, and frogs too. The same math also seems to hold for plants, at least for their vessels and metabolic rates.[7] No animal escapes these associations; no animal escapes the math of West and his colleagues.[8] "Sometimes," West said once, "I look out at nature and think, Everything here is obeying my conjecture." It's as though he were the wizard behind the screen, calling each animal into existence.

Perhaps not surprisingly, West's laws are subject to great argument among biologists. Some biologists feel they have simpler models that work just as well. Others debate whether he has explained why these patterns among bodies or cities exist; they argue over the extent to which West has sacrificed the interesting details of the living world, the important particulars, in search of, perhaps, overly simplified generalizations. They debate the slopes of relationships. West's laws are undeniably coarse and sweeping, but it is their sweep that makes them interesting. These laws can even account for changes in our bodies as we age and our bodies get bigger. The hearts of babies beat far faster than those of adults (the heart rate of a newborn is a supercharged 185 beats per minute). The heart must do this to keep the baby's body warm (within a species, this relationship has its limits. Obese individuals do not have slower hearts, even though it is much easier for obese people to keep themselves warm).

But West can predict more than how a body works; he can also predict how it will fail to work. This was where cities and bodies diverged; cities fail, sure, but they do not appear to have a natural longevity per se. Not so bodies. Bodies fail on some sort of schedule, and West thinks he can explain when and why. These predictions won't lead anyone to the fountain of youth, but they may explain why the fountain of youth is desired in the first place. As West put it in John Whitfield's book *In the Beat of a Heart*, "If biology is to be a real science, you ought to have a theory that can predict why we live 100 years." He then proceeded to extend his math and physics and insights a little further still.

The Pace of Life

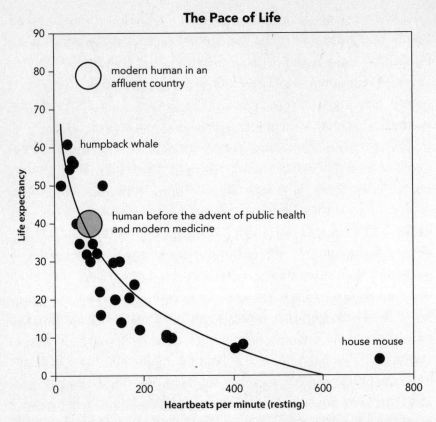

The life expectancy of mammal species as a function of their resting heart rates. Mammals (and birds, not shown) with high resting heart rates tend to live fewer years than mammals with lower resting heart rates; nearly all mammals get about a billion heartbeats. Historically, humans are no different, but with public health and modern medicine we have escaped these constraints and, in doing so, live out about a billion extra beats.

We tend to take it for granted that the life expectancies of different organisms are different. We talk about "human years," for example, as compared to "dog years." But it is not obvious that longevity should vary. Mammal cells, after all, are essentially the same. Yet, as West found in the literature, with a pen and paper, you can graph the relationship between heart rate and the maximum longevity of a species, or how long individuals under the best of circumstances live. So far, *nearly* without exception, the longevity of a species is predicted by its heart rate. The relationship is a straight line on a semi-log plot. Species whose hearts beat faster live fewer years. An Etruscan shrew lives one year, a blue whale more than a hundred years. In these lives, their hearts beat about the same number of times: one billion. West thinks this longevity follows from the sizes of bodies, which in turn relates to the rate of hearts. From West's perspective, the rate of the human heart determines the maximum sustainable metabolic rate and activity level of each cell and, within each cell, each mitochondrion. Indirectly, then, West thinks, body size and heart rate are a measure of how much wear there is on each tiny and large part of the body.[9] This wear influences everything, from the buildup of atherosclerosis in arteries to the ability of the body to continue to feed beneficial microbes.

Ultimately, in accordance with West's laws, all wild species get a maximum of about a billion heartbeats.[10] The only difference is the time over which the beats occur. A shrew uses its heartbeats quickly; a whale savors them, lounging as blood flows from its big muscle out to its brain and its long, long tail. But could our fates, human fates, really be predicated simply on the number of beats of our hearts, and if so, what do such predictions, leveraged from comparisons among species on separate evolutionary trajectories, say about our fates?

Of course, a truck can hit an animal. Animals can be struck by lightning. They can be eaten. Many calamities befall them (and us), as they always have. Heart rates, though, seem tied to the maximum life span an individual in a population of a species can expect to live if everything goes right. But what about humans? Historically, the life expectancy of a human in a small population was around forty years, somewhere between the shrew's life span and the whale's; it falls right where it would be expected to, given the human heart rate and body size. Historically, humans too got about one billion beats.

If heart rate really does affect an organism's fate so strongly, one can gin up a few hypotheses in need of testing, particularly if one begins by assuming (as most scientists have) that heartbeats actually use up the body in some way or another, breaking it down in ways it cannot repair. The first, most obvious prediction—if heart rate really influences our longevity—would be that organisms who can slow down their hearts, whether through hibernation or torpor, should live lives longer than would be expected given their average heart rate in their active months. In slowing themselves, they should get a number of days or even years of free ride.

Animal hearts slow to differing degrees when they need to. Blue-whale heart rates slow when the whales dive (to as few as three beats per minute). Blue-throated-hummingbird hearts beat up to twelve hundred times per minute when they are flying but drop to thirty beats per minute when they are asleep. More conspicuously, the hearts of many mammals slow when they hibernate.

We tend to think of bears when we think of hibernation, but research has shown that bears are not actually true hibernators.[11] They slow their hearts, but their bodies remain warm, so they are somewhat awake the entire winter, ready to take advantage of a good day. Their hearts slow the way yours might if you were practicing yoga. In one study in India in which one group of participants practiced yoga for ten days and another group did not, those who did not saw no change in their resting heart rates, while those who had practiced yoga saw their resting heart rates decline by about eleven beats per minute, even though they had done it for only ten days, and even if they hated yoga.[12] Yoga practitioners, in other words, enjoy a sort of permanent hibernation not unlike that which yields sleepy (but testy) bears.[13]

Both bears and yogis might be expected to have longer life expectancies than they would if their heart rates were high year-round, but for a more extreme case, we need to turn to the ground-hog. Groundhogs (*Marmota monax*)—also known as whistle pigs or woodchucks—live in grassy areas throughout North America and resemble inflated squirrels.[14] Groundhogs are conspicuously fat. They can weigh as much as thirty pounds, and they spend their days walking slowly up hills and down dales, eating.

Groundhogs have the sort of body type one might imagine could exist only on an island where there were no predators. They remind me of miniature elephants, giant tortoises, and other unfathomable beasts of distant, predator-free lands.[15] Groundhogs, though, have an escape plan. They dig deep warrens in the ground into which they plunge at anything remotely resembling danger. Each warren has four or five separate entrances. This is what has saved them from many predators over the several million years of their species' existence. The predator lunges, runs, or jumps, and the groundhogs dive. The groundhog is one of the silliest-looking mammals in the world, but it proved important in one area of the study of the heart.

The big challenge for groundhogs is winter, when their food disappears beneath the snow. Larger herbivores deal with such scarcity by moving: caribou migrate and moose stroll and amble and reach higher into the trees. The groundhog has no such luxury; it is chained to the safety of its hole and is too fat to get very far. It must hibernate, and it must hibernate well enough that it is not forced to vacate its safe residence to look for food that is just not around. Hibernation in the style of the bear is relatively easy. Bears eat until they are bulbously round and then camp out in a cave and let the fat burn off. If they get too hungry, they go out and munch something.

Humans could almost hibernate bear-style, but not groundhog-style. The groundhog goes underground in the late fall and, when it does, turns its body temperature down so far that its core temperature decreases ten degrees centigrade, from 38 degrees C down to 10 degrees C, so far that it doesn't really need to be that fat to survive the winter, so far that its heartbeat slows dramatically. Its metabolism is 1 percent of normal.[16] If the allometricians are right, the groundhog should have a maximum life expectancy that is much longer than might be predicted given its size and normal resting heart rate. It should get bonus years in proportion to the number of heartbeats it saves thanks to its long winter's rest.

The summertime heart rate of the groundhog is an ordinary eighty-nine beats a minute, slightly faster than yours or mine. But the wintertime heart rate of the groundhog is just ten beats per minute. And, indeed, the groundhog lives 30 percent longer than we would expect it to based on its summer heart rate. Its winter doze does it good, allowing it to space out its billion beats over a longer period.

It turns out the groundhog is not alone. All organisms that hibernate add years to their lives, and the more fully they hibernate, the more years they get. This even works when we look at a group of organisms in which some but not all hibernate. Bats that hibernate, for example, live longer than those that do not.

Nor is hibernation the only slowed-time behavior that has an impact. Hummingbird hearts beat more than a thousand beats per minute when they are flying, but when they land, their heart rates decline to just sixty beats a minute. At night, when they sleep, their hearts nearly stop. As a consequence, hummingbirds live much longer than shrews, even though both of their hearts can beat very, very fast; shrews never slow down. It really does look as though, if you know how many heartbeats an animal has—no matter when or in what bursts—you *can* predict how long it will live. Essentially all tests of this suggest a natural limit to the number of heartbeats an animal gets. All wild animals get about a billion beats; they just get to use them slow (if they are big or hibernate) or fast. Cats may have nine lives, but they too get just the standard number of heartbeats, a billion, give or take, and no more.

West's law linking mitochondria, metabolism, body size, heart rate, and longevity has the potential to lead to many insights about the limits of our bodies and hearts. But the idea of studying hibernating mammals and their slow hearts to extend human lives is not totally new. Long before the modern research on heart rate and life expectancy, there was a study on heart rate in groundhogs that yielded what seemed like one of medicine's most penetrating insights, an insight about the potential to slow our own hearts down and, in doing so, extend our lives.

In the late 1940s, W. G. Bigelow was a rather ordinary surgeon and researcher at the University of Toronto. He was interested in the heart, but not unusually so. Like many, he wanted to figure out a way to conduct successful heart surgeries. In the 1940s, the chest had been opened and the heart operated on, but there was that limit to what was possible, that old three-minute problem. At about the time that Gibbon went to war, leaving his heart-lung machine at home, Bigelow stumbled on another approach. He had been puzzling over how to better operate on the heart when, fortuitously, a man came into his office suffering from frostbite. Bigelow noted that the man's heart was beating more slowly than it normally would, and yet he was alive. Seeing this, Bigelow began to think about the ways in which bodies could be cooled, and it led him to the controversial theory that human bodies could be cooled in order to allow heart surgeries to be more easily performed or even, more generally, to extend lives.

When the body is cooled, Bigelow had come to know, the heart slows, and the body's demands for oxygen also decline, for the simple reason that cells in the body are metabolizing more slowly, requiring less oxygen. Beginning in 1947, Bigelow did experiments on cold dogs. In one such experiment, thirty-nine dogs (one suspects a fortieth didn't make the cut) were cooled to 20 degrees C. It was difficult work because the dogs' shivering made it hard to cool them all the way, but Bigelow did his best. Then, in each dog, Bigelow blocked the flow of blood back to the heart, essentially shutting the heart off. He kept it blocked for fifteen minutes, five times longer than a dog's body should be able to go without the beating of the heart and its froth of oxygen, at least at normal body temperature. Once blood flow was restored, 51 percent of the dogs survived; they survived with far less oxygen than they normally needed from the heart because each of their cells had slowed down.

In theory, Bigelow's results meant that cooled humans could have their hearts stopped for more than ten minutes and still possibly be resuscitated, though with only a 50 percent survival rate, the procedure was still too nascent to be applied to humans. Yet it was all very exciting (albeit not for the dogs). Bigelow announced his idea at an American Surgical Association meeting in Denver, Colorado, in 1950.

Almost immediately, eager young scientists, led by John Lewis, followed up on Bigelow's idea and developed an entire subfield in which bodies would be cooled in order to facilitate heart surgeries. The earliest efforts, the first of which was an attempt to repair a congenital heart defect, were made so quickly after Bigelow's talk that he had not even had time to publish his findings before the surgeries were complete. Lewis, as Bigelow would say, "broke the ice." The 50 percent survival of Bigelow's dogs had not concerned Lewis.[17] Cooling bodies worked; the first patient survived open-heart surgery. She had been placed on a bed of ice, and her heart was stopped for ten minutes. The historic, three-minute barrier was surpassed, and with it, the possibility of far more ambitious surgeries had opened up. Heart-lung machines would eventually replace cooling for many surgeries, but today the two procedures are sometimes used in concert. Bigelow's insight was fundamental and correct, but for him, it was still just a sidestep en route to an even bigger lesson.

After his initial work cooling bodies for surgeries, work others seized upon, Bigelow himself chose another trajectory, one that brought him ridicule for the rest of his life. He had noticed that while human bodies could be cooled and their hearts slowed, their cooling was different from that seen in hibernating animals. Hibernating animals seemed to be able to cool themselves and slow their heartbeats without external help, such as ice. They cooled of their own physiological volition. When they did, they did not experience shivering, and their bodies did not fight the process (a big problem with both dogs and humans). This fact, Bigelow thought, could be used in human surgery and maybe even to extend human lives. The body temperatures of groundhogs could be lowered to 3 to 5 degrees C in the lab, after which their hearts could be disconnected for as long as two hours with zero deaths. While many of the dogs that Bigelow had cooled died, the cooled groundhogs almost never did. When he finally wrote up his 1950 paper (in 1953), Bigelow noted, "A greater knowledge of hibernation may yield useful information on this problem."[18] Bigelow began to think that hibernating animals had evolved a compound—he ambitiously gave it a name, *hibernin*, even before he discovered it—that triggered the cooling of the body and the slowing of the heart and metabolism. This compound, he thought, allowed these animals to hibernate but also extended their lives. Bigelow hoped to find this compound and use it to extend human lives, whether while surgery took place or perhaps via some other means.

To find this chemical, Bigelow would study hundreds of groundhogs, devoting ten years of his life. He studied them under the conditions in which they were likely to be producing the mythical hibernin: outdoors, in the winter. This is not an easy thing to do. The warrens go deep and are complex, and groundhogs can react negatively to being yanked out of their holes. Bigelow persevered. He crawled into tunnels on his belly. He pulled animals out. He raised furry little groundhog babies. He set up the biggest groundhog study in history, probably never to be repeated. At its peak, his groundhog facility housed more than four hundred animals, all tunneling, eating, and hibernating just north of Toronto.

Then, after six long years of research, there was a moment of elation. Newspapers were abuzz in the 1950s with the idea that, according to Bigelow, there was a "strange brown-colored fatty tissue" around the groundhog heart that, if removed, made the groundhogs much more sensitive to cold. Maybe this was the magic stuff of a hibernation gland? Bigelow injected the substance into other animals—both guinea pigs and rats—and successfully cooled their body temperatures to 5 degrees C. Animals without the injections could be cooled to only 14 degrees C.

Bigelow was overwhelmed with excitement. He suspended his surgical practice to focus on the groundhogs and hibernin. He even injected the substance into two human patients, both of whom survived very low temperatures but seemed drunk. This was an odd finding and, it would turn out, a telling one. Bigelow submitted a patent for hibernin. It was then that he realized something had gone terribly wrong: The patent was rejected. The substance he had submitted a patent for had already been discovered and patented by someone else, and it was not a miracle compound. It was a plasticizer, a compound used in laboratory tubing, safety goggles, and other laboratory plastics to keep them pliable. Some plastic from the lab equipment appeared to have gotten into the samples. It was that compound that he had been injecting—a compound whose active component was butyl alcohol (hence the drunken behavior). With this, Bigelow became known as the man who had inspired the use of cold in heart surgery only to lose himself among the groundhogs.[19]

But science is complex. One decade's loss is another's gain. Bigelow eventually retired, but not before planting a few seeds. The field he inspired progressed, as young scientists continued to study hibernation, metabolism, and heart rate. Scientists abandoned the quest for something so simple as hibernin but not the attempt to figure out what makes groundhogs and their kin so special. Then, in 2012, a team at the University of Alaska, led by Tulasi Jinka and Kelly Drew, discovered hibernin. Working with ground squirrels, Jinka and Drew had been trying to answer the same question Bigelow posed, but they had the advantages of the major advances in scientific technology that had occurred in the past fifty years. Jinka and Drew isolated a compound that, when released in the ground squirrels' bodies, caused the animals to hibernate. Jinka and Drew are now able to perform an amazing trick with this compound. They wake hibernating ground squirrels up. Once awake, the ground squirrels stay awake, thinking they have found spring, until Jinka and Drew give them the magic compound, which causes them to go right back to sleep. The compound is called adenosine, not hibernin, and yet it does exactly what Bigelow had suspected hibernin might.

Could adenosine be used to slow the metabolism of humans? Could it be used to extend lives or put people, Rip van Winkle–style, into suspended sleep to wait for new cures? Maybe. The good news is we actually know a fair bit about adenosine. It is used clinically to slow heart rates in certain kinds of dangerously fast heart rhythms. In nature, it induces hibernation in ground squirrels, but interestingly, it does so only during the winter (and it doesn't do so if they have been given caffeine).[20] It appears that during the winter, the receptors to which adenosine binds become more receptive or responsive, making hibernation possible. This may mean that in order to use adenosine more effectively in humans, we need to figure out how to use both it and its receptors. More remains to be discovered, as always, and yet a big corner has been turned. Sadly, all of this was realized seven years after Bigelow's death, in 2005. He had been on the right track. He just needed more time.

As Bigelow had intuited, hibernators have something to reveal, but so do many other wild species. The past few years have produced a richness of examples of what can be learned from the hearts and blood of other species. Burmese pythons, for instance, have hearts that wax and wane with meals. Could lessons from these pythons be used to help us figure out how to regrow parts of human hearts? Yes. The groundhog, ground squirrel, and python are just three species from which we can learn, three out of millions. Further study of the hearts of other species will teach us about the limits (or maybe even advantages) of our own.

A big lesson we've already learned from the experiments of Bigelow and from Geoffrey West's laws of scaling is that the life expectancy of an animal is, on average, equivalent to one billion heartbeats. The beats can be slow, like a tortoise's. Or fast, like a shrew's. Or in pulses of fast and slow, like a hibernating animal's. But that is all most animals get.

Still, exceptions can be found. Some wild animals live longer than would be expected given their heart rates. In some cases, this appears to be related to how their mitochondria work and just how, at the finest scale, they wear down. Then there are humans. When Bigelow began his work, in the 1940s, humans lived lives of just over a billion beats. Now, in the United States and much of the developed world, the average human gets about 2.5 billion heart-beats, 1.5 billion of which are extra, bonus years thanks to the successes of modern public health and medicine. One perspective on this change is simply that we are extending our lives; another is that we are living out more heartbeats than any species on Earth ever has. In terms of an internal, biological clock, each of us lives, on average, two full lives.

Our bodies break with the same frequency as those of other species, but we can mend them before they fall apart (as with the treatment of congenital diseases) or, better yet, find ways to prevent them from breaking in the first place (as in the case of statins). If you are lucky enough to be born in a fortunate part of a fortunate country, if you get sick, your body can be tended to. Much of the success, at least sixty years' worth, is due to improvements in how we care for our hearts.

People like Bigelow, Gibbon, and all the others have given each of us a chance at a second life, a billion and a half heartbeats with which to do as we please. Taussig used hers to keep making discoveries. Bigelow used his with the groundhogs. Time will tell what Geoffrey West will do, though it seems likely he will keep walking to his office, struggling with life's data. Time will also tell what the rest of us do with our beats, those billion gifts, moments in which to change the world or just sit back and admire it.

Postscript: The Future Science of the Heart

I have a strange job. I get to wake up and study the species that live on humans and around humans and how they affect human lives. This is work I do with the public. Who better, after all, to study bodies and homes than the people who inhabit them? In studying the species around us, my colleagues and I have discovered a large number of things no one ever knew were there. One day, we find fifty unnamed bacterial species in someone's belly button. Another day, we find a totally new kind of animal living in the pores on a teacher's face. In houses, we have discovered a wasp that no one can name; what we know of it, we have garnered based on what is known of its relatives—namely, that it likely makes its living by laying eggs in the body of some other animal, where they then develop and eat the other animal, the host, from the inside out. This wasp is very common in homes but goes unnoticed. Elsewhere, we have discovered a thumb-size cricket that has spread basement to basement across North America without note. Everyone thought someone else knew it was there.

I mention all of this because there has been a consistent lesson for me in studying these species. It is a lesson that rings out loudly and unambiguously every time I go to work: we humans are far more ignorant than we imagine. I wrote a whole book about this reality earlier in my career, and still it strikes me. What I didn't realize when I was writing that book was that the more focused we are on our daily lives, the more likely we are to miss big discoveries. You expect big discoveries in the rain forest and so look for them. In your house, they run under your feet and you miss them, or perhaps you figure that someone knows what they are even if you do not.

Here, at the end of the book, I'd like to be able to tell the future of our hearts, and the one thing I am most sure of is our ignorance and, hence, the future's uncertainty. We know far less about hearts than doctors, scientists, and everyone else thinks. Helen Taussig posited that out there, flying near your house, there might be a bird with two hearts beating in unison (or maybe not beating in unison). The existence of such a bird seems unlikely, and yet we can't really rule it out. By the same measure, we can't rule out the existence of many kinds of hearts far different from those we know, or features of our own hearts that we haven't yet understood. Generation after generation of scientists have assumed the body was, if not perfectly understood, nearly so. They were wrong. Our generation will be no different.

Part of dealing with our ignorance is humility. But there are also practical ways of improving the odds of making big new discoveries. When it comes to understanding how our hearts work and fail to work, it is a good idea to try to understand the hearts of other species. Nissi and Ajit Varki are starting to compare chimpanzee hearts and human hearts, but they have really just begun. We know very little of the other apes, much less the other primates in general. Given how much light a modest understanding of chimpanzee hearts has shed on the understanding of our own hearts, it is easy to imagine the light from other apes will also be clarifying. And it isn't just the apes. There are more than five thousand mammal species on Earth, each with a different type of heart. There are around twelve thousand bird species. And then the fish; there are tens of thousands of kinds of fish, and millions of insects. Each of these species carries some lesson to be understood.

But there is more. Cyclosporine, it is worth remembering, comes from a fungus that immobilizes the immune systems of beetles. Statins come from fungi, a product of their wars against other microbes. Antibiotics that kill off the pathogen that causes rheumatic fever, bane of the heart, come from other fungi fighting other bacteria. In the millions of wild species are millions of answers.

Predicting the future of medicine may well be more difficult than predicting the future of discovery. It depends not only on science and technology but also on governments, policy, and culture. Bless those who imagine they know the future of policy and culture. The past, though, offers some insight. For one, we do not have to squint too hard to notice that in the story of the heart, there are repeated cycles of hubris, the appearance of accomplishment, and then prolonged or even ultimate failure.[1] I'll speculate that, similarly, part of what seems like the light of progress today is illusion; humans are drawn to the bright light of accomplishment more strongly than to the subtler candle of reason.

The trick is distinguishing progress from illusion. Brighter minds than mine have tried and failed. Today, one hope on the horizon involves stem cells. In labs around the world, researchers are racing to use stems cells, sprinkling them on the heart or injecting them into the heart to regenerate its muscles. Stem cells are all-powerful cells; they can be anything. These new treatments seek to convince these cells to become part of the heart.

In petri dishes, entire beating heartlike blobs of cells have been created out of heart cells left to grow on a scaffold. Some researchers are now talking of making hearts in the lab out of human cells. We could, they say, grow thousands of hearts. We could grow parts in general. We could grow ourselves into eternal lives. No one is so bold anymore as to mention eternity or immortality, but quite a few folks are talking about extending lives decades more on top of what they are today.

Right now, the ability to get stem cells to regrow heart tissue is as exciting as the first heart transplants done by Shumway and Lower on dogs in the 1960s were. Just as with the dogs, it is immediately clear what the intent is, what the future might be. In fact, clinical trials are already going on around the world. More than a thousand people have had stem cells released into their hearts. So far, this sprinkling of new, potent cells has not helped the patients. It turns out that, like the ladybugs you release into your garden, stem cells have a wandering jones. Release them into the heart and they wash away, travel elsewhere in the body. Now folks are using mice to try out devices that slowly release the stem cells. In mice, these seem to work better. We wait. The field is full of both hope and contention about these new approaches.[2]

Hundreds of researchers are working on stem cells and hearts, and so progress might happen quickly. But these hundreds, in addition to giving me hope, also remind me of the other thing I can predict, this time with certainty. Although society invests heavily in projects that appear likely to result in immediate medical treatments, it invests far less in research that gives context to the need for such treatments. Only a handful of scholars study why our hearts are prone to atherosclerosis in the first place. No one is following up on the hearts of Taussig's birds. Essentially, no work is being done on how and why human hearts might differ from one population to the next around the world (they almost certainly differ). We have left the study of who we are and why to a handful of diligent folks whose work has the potential to fundamentally rearrange our understanding of our bodies and yet is little known and poorly funded.

I don't think the relative neglect of basic biological studies of the heart (or any other organ) will change, and so, rather than bemoan the situation, I have a recommendation. If you are a young person and want to grow up to understand the heart, to make a great discovery the elegance of which people can only begin to contemplate now, study the heart's ecology and evolution. Study hearts around the world. Study the hearts of frogs and turtles and, especially, as it turns out, snakes. Study the biology of wild nature in general. This course of action is not likely to make you rich or famous, but it may well lead to a moment, somewhere in a small lab, or better yet in the middle of jungle, in which you suddenly realize something about the heart that no one else, no one in the history of science or society, had even considered. That thrill will be worth it, worth some sacrifice. That thrill will give you goose bumps that never quite go away.

But back to the snakes. I love snakes. I think there is a convincing case to be made that they have shaped our evolution, so great was their historical impact on human life and death. As a result, a kind of wild fear of snakes is come by honestly. Yet it does not take much inspection of a snake to realize that magic dwells in its armless existence. Arms and legs took hundreds of millions of years to evolve, and the snakes, who once had them, decided to do without. They slither. Their backs bend in remarkable shapes, thanks to a multitude of vertebrae. Their jaws expand to envelop prey bigger than their heads. Their tongues reach out to gather scent in the air and then return it to a special organ in which those scents can be stored and interpreted. Snakes are special, and this specialness, it has recently been discovered, extends to their hearts.

Snakes have three-chambered hearts—two atria and one ventricle. The heart of a snake works similarly to that of a human with a ventricular-septal defect. Each time the ventricle pumps, some blood goes to the lungs, some to the body. It is a sloppy business, but functional.

When snakes are dissected, even snakes of a single species, their hearts are sometimes found to be enormous, and sometimes small.

It was noted that the heart seemed bigger after the snake had eaten a meal. Appearances and anecdotes are not to be trusted, though. The idea of the snake's heart growing during digestion is clearly ridiculous.

Human hearts can change some in size, but they do it slowly. When the heart is injured, some cells divide and others expand. The heart remodels itself, although to a very modest extent. Bigger changes occur during pregnancy, when a woman's heart expands to pump blood through two bodies; during development, when a newborn's heart grows rapidly; and during prolonged exercise, when the heart expands to move blood. But even these beneficial changes are small, a 10 or 20 percent increase in size at most, and the higher percent only a freak-show potentiality.

There is no model snake for biological research, but Burmese pythons (*Python molurus bivittatus*) seem to have been studied as well as any. These pythons, which can grow as long as nineteen feet, can go an entire year without feeding. They live slow lives, except for those voracious moments when they don't, those moments in which they seize prey, sometimes as big as they are. The muscles of Burmese pythons speed up when they attack prey, but the real shift comes during digestion. Steven Secor, a professor at the University of Alabama, Tuscaloosa (and a reptile chaser by native inclination), and his former adviser Jared Diamond, of *Guns, Germs, and Steel* fame, discovered that the metabolic rate of Burmese pythons during digestion is forty-four-fold what it is when they are hungry. It might be advantageous, then, for these big snakes to be able to expand their hearts after they eat. Having eaten, they need more blood and oxygen for digestion. Having eaten, they have more amino acids, triglycerides, and free fatty acids to move around. Having eaten, their bodies have a great deal to do. Sure enough, a recent series of studies by Secor and colleagues has shown that forty-eight to seventy-two hours after a snake eats, its heart expands by 40 percent (its liver, intestines, and kidneys also expand). With the increase in heart size, the snake's output of blood goes up fivefold.[3]

The enlarging of the python heart appears to be due to increases in the size of individual cells. One of the things that changes in the python's heart and blood during the expansion of these cells is the number of fatty acids in the blood; they go up dramatically (as much as fiftyfold), which appears to trigger the growth of the heart cells. All of this is the obscure business of snake biologists such as Stephen Secor, merely a lovely story about the eccentricities of our legless friends. Or it would be. But one of the teams of researchers studying the python heart, a team collaborating with Secor, decided to inject the same cocktail of fatty acids that are found in snake blood into the hearts of lab mice. The heart cells in those mice grew,[4] as did rat-heart cells in culture. Our own heart cells, it seems clear, might too.

The ability to make heart cells grow larger might hold therapeutic value. One of the problems with diseased human hearts is that they become hypertrophic (overly large), and in part, this hypertrophy is due to the expansion of cells. Understanding how the python-fat elixir makes cells grow might allow scientists to better understand such growth in diseased human hearts and, perhaps, find a way to prevent it. It might also offer some benefit in cases of atrophy, in which the growth of heart cells would be helpful. Compounds from other reptiles are already being used medically. A compound in Gila monster saliva, for example, is the active ingredient in the diabetes drug Byetta. There is much to learn, and some things we will learn from aggressive study of humans and from big medical experiments, but bigger discoveries dwell in the bodies of snakes and the millions of species we don't yet understand. The light of our inquiry remains humble relative to the grandeur of existence.

As for me, one of the great joys of being a scientist and a writer is that when I realize we don't know very much about something, I can just go study it. In the next months, I may begin a study of the bacteria and viruses found in heart tissue. We know little about the microbes of the heart other than that they are almost undoubtedly there, dividing, thriving, doing whatever they do. Are they the same ones found in other primates? We don't know. Perhaps we will need to study the hearts of chimpanzees or gorillas. Perhaps the viruses and bacteria in the heart come from the gut. Whatever we find, I predict it will be fascinating and new. Here, I am not just guessing; I am betting on the prediction that always seems to hold—namely, that we find something new each time we look, especially when we consider a great and poorly trammeled wilderness such as ourselves.

ACKNOWLEDGMENTS

Later in his life, my great-grandfather, who was active in his Methodist church in Greenville, Mississippi, was asked to comment on the history of that church. I recently discovered his response in a box of old family letters. He wrote, "In commenting upon the history of the Methodist church in Greenville, I would be remiss if I did not comment upon the history of the Methodist church more generally. In commenting upon the history of the Methodist church more generally it behooves me to speak to the history of Christianity and, of course, no discussion of the history of Christianity would be complete if it did not include a discussion of the religions of the world."

I come from a long line of people who, when telling a story, back way up to the beginning. For that same tendency in my own writing, I thank my great-grandfather. I thank my grandfather, his son, for the joy I find in exploring everything. I hope you, the reader, find some of his joy in this book. I thank my grandmother who grew up living in the building at the University of Mississippi where the biggest telescope in the world was supposed to go except that the telescope was built in the North and not yet delivered when the Civil War started (and so never delivered). Instead of being able to see out into space, she grew up listening to Faulkner tell stories on her front porch. As a result, she could instead see far into the universe of people, and for that I thank her. My mother allowed me to share the story of her heart. My dad read the book and reminded me when I was being too much of a scientist. Thanks to them both also for loving everything I write, even early on in my writing when, in retrospect, it is clear that when they told me that, they were lying.

My wife gave me the guts to tackle a topic as central as the heart. She also put up with the consequences of that confidence: thousands of conversations about the characters in this book (and an even larger number that never made it in), characters that, while fascinating, are a little much for dinner conversation every week for two years. Thank you, Monica, for your push, forbearance, wisdom, keen editorial eye, all the funny parts of the book, and everything else. My kids too listened to stories about the heart; Lula and August know more about blood vessels than any eight- and four-year-old really need to.

Many people have read parts of this book or responded to interview requests.

Bill Parker (the man who, in my last book, discovered the true function of the human appendix) read the book and added to it his special bit of magical brilliance. Colleen Farmer, Keith Myles, Will Kimler, Abell Assam, Ajit Varki, Nissi Varki, Kathie Hodge, Mariano Vázquez, Mohammadali M. Shoja, Nick Haddad, Stephen Secor, Geoffrey Donovan, Sarah Tracy, Herbert Cohn, Chris Gould, George Forssmann, Anne Murphy, Jie Jack Li, Kymberleigh Romano, Mizuki Takahashi, Harry Greene, Andrew Latimer, James Waters, Pajaro Morales, and Mette Olufsen all read sections of the book, in nearly all cases more than I asked them to. Dr. Bill Haynos read the book at the last minute, frantically, late at night, even when he had to get up early the next morning to look at his patients' hearts. Thank you, Bill. Thank you also to Steve Jordan for the introduction and for being a supporter of this arcane work of turning words into books. Amanda Moon and T. J. Kellerer provided thoughts on what worked with the concept of the book that were very useful. Marko Pecaravic provided a balcony in Croatia on which I wrote part of this book. Michelle Trautwein and Ari Lit listened to these heart stories again and again, pretending the stories were interesting even before they were. Steve Frank listened to these stories while walking around his neighborhood late at night.

John Parsley and Malin von Euler-Hogan wielded an ax where necessary and a fine carving tool everywhere else. Thank you, John and Malin, for your patience, vision, and extraordinarily clear thinking. Thank you also for finding Tracy Roe, a copyeditor who's also a physician. Who knew such people existed? Thank you, Tracy. Victoria Pryor helped with everything, even when she was busy, even when it wasn't her job. Every part of the book is better for it.

And then I also need to thank my lab. Thank you, Holly Menninger, Lea Shell, Clint Penick, De Anna Beasley, Amy Savage, Amanda Traud, Magdalena Sorger, MJ Epps, and everyone else for being wonderfully patient when I disappeared into a coffee shop, library, or basement for days on end to write. Thank you especially, Emily and Megan, for reminding me, with your stories, of the ordinary urgency of heart problems. And thank you, lab, because, as you all know from experience, when I write a book, the mysteries revealed in a book come back to the lab. How could I not go back to the lab and study those things that appear unknown and yet knowable? And so thanks in advance for everything you help with in trying to understand the mysteries of the heart, mysteries that began, as my great-grandfather would have pointed out, thousands of years ago and yet, if we have some luck, might end in our third-floor lab (only, of course, to spawn new mysteries).

Endnotes, References, and a Few Anecdotes

1. The Bar Fight That Precipitated the Dawn of Heart Surgery

1. An early perspective on Williams's story can be found in W. M. Cobb, "Daniel Hale Williams—Pioneer and Innovator," *Journal of the National Medical Association* 36 (1944): 158. For perspective on Williams's broader career, see W. K. Beatty, "Daniel Hale Williams: Innovative Surgeon, Educator and Hospital Administrator," *Chest* 60 (1971): 175–82.
2. Provident Hospital is described by some as the very first interracial hospital in the United States. See W. M. Cobb, *The First Negro Medical Society* (Washington, DC: Associated Publishers, 1939). Whether or not it was first, it was certainly among the first.
3. Emma Reynolds would go on to obtain her MD at the Woman's Medical College of Chicago in 1895. She then moved to Waco, Texas, and, later, New Orleans, where she would practice medicine until her death in 1917.
4. Details about Cornish's night in the bar come from S. Cohn, *It Happened in Chicago* (Guilford, CT: Morris Book Publishing, 2009).
5. Also lacking at this point were antibiotics, a heart-lung machine, blood transfusions, and intravenous anesthesia, to name just a few items.
6. Sounds, like sights, tell stories. We have all heard someone trying to replicate a sound his car makes: "It goes *flappa, flappa, flappa*," or "It's like *ticka, tank, ticka, tank, squeee!*" Our bodies too tell stories of our well-being. They tell them via their appearances and odors but also via their sounds, which were once used extensively to make diagnoses. Doctors would actually put an ear against the patient's chest to listen to his heart. Then, in 1816, a Parisian named René-Théophile-Hyacinthe Laennec invented the stethoscope. Laennec had been watching children play a game in which they scratched a long, hollow, flutelike stick on one end with a pin and

331

then listened for the scratch at the other end. This, he thought, might also allow him to listen to hearts. Laennec promptly went back to his office to experiment with a kind of flute, a long tube that he might hold to the chest. With this device, the first stethoscope, many of the mechanical problems of the heart could be differentiated by sound. An audio archive is now available to teach new doctors the art of listening to the heart's problems through stethoscopes, though most doctors aren't skilled in listening to the heart; technology has allowed the practice to fall by the wayside.

7. For these estimates see J. L. Halperin and R. Levine, *Bypass* (New York: Times Books, 1985).

8. D. H. Williams, "Stab Wound of the Heart and Pericardium—Suture of the Pericardium—Recovery—Patient Alive Three Years Afterward," *Medical Record* 51 (1897): 439.

9. H. C. Dalton, "Report of a Case of Stab-Wound of the Pericardium, Terminating in Recovery After Resection of a Rib and Suture of the Pericardium," *Annals of Surgery* 21 (1895): 148.

10. L. Rehn, "On Penetrating Cardiac Injuries and Cardiac Suturing," *Archiv für klinische Chirurgie* 55 (1897): 315.

11. "Heartbeats," *Time* 1 (1923).

2. The Prince of the Heart

1. And just as often of the terribleness of his mother, who, he claimed, "bit the servants."

2. One sometimes needs to be a bit circumspect in considering Galen's biography. He wrote most of it. In fact, while he was alive, he wrote most of what was written in the Roman Empire having to do with medicine—millions of words.

3. One place we do find pan-cultural knowledge of the heart is on the dinner table. Hearts are (and, one presumes, have long been) cooked in cultures around the world. As Steven Vogel, in his excellent book *Vital Circuits*, points out, hearts are not easy to cook because they are full of collagen (the original source of glue, which Vogel also notes). Collagen is chewy and hard to eat. The best way to deal with collagen is to break it down molecularly, which can be done by leaving the collagen in acidic solutions such as vinegar, lemon juice, or tomato sauce. Then, of course, add some red pepper, cloves, cumin, and the like for flavor, and cook.

4. It was for this reason that in some groups, the kidney fat was the part of the body most relished in acts of ritual cannibalism. In eating the kidney fat, one consumed a soul.

5. The Egyptians actually had two separate words for the heart, one for the spiritual heart, *ib*, and one for the physical organ, *haty*.

6. R. Van Praagh and S. Van Praagh, "Aristotle's 'Triventricular' Heart and the Relevant Early History of the Cardiovascular System," *Chest* 84 (1983): 462–68.

7. Herophilus is also sometimes described as the most prolific vivisectionist in history. Tertullian describes Herophilus as having dissected six hundred living criminals, though if this was really true, it seems as though Herophilus would have figured out more about how the heart works, having seen so many beating their exposed, last beats.

8. In a rare moment of modesty, he confessed to the limits of his knowledge, admitting to never having dissected an ant, a louse, or a flea.

9. Because only a portion of what Galen wrote has survived, much of it in secondhand translations, it is sometimes difficult to disentangle exactly what he believed.

10. Nor were these treatments the extent of Galen's lingering influence. For example, the practice of dissecting animal bodies in anatomy classes in order to reveal the truths in biology textbooks is simply revisiting the ancient Galenic practice.

3. When Art Reinvented Science

1. S. J. Martins, "Leonardo da Vinci and the First Hemodynamic Observations," *Revista Portuguesa de Cardiologia* 271 (2008): 243–72.

2. All of that said, da Vinci did have preconceived notions as to the causes of death in old age. In scholar Kenneth Keele's read, what da Vinci imagined he was looking for was whatever prevented the movement of "vital heat and humors" through the body. In other words, he started off looking for blockages of one kind or another. Lack of movement was like stagnant water that putrefied. Here, then, the Galenic perspective on the body (a perspective in which vital heat and its movement was key) actually predisposed da Vinci to see the truth. It is also interesting to note that da Vinci's analogy of stagnant water to blocked blood flow is, in its way, quite apt. Stagnant water is devoid of oxygen, and blocked blood flow leads to the inability of oxygen to reach the body. See K. D. Keele, "Leonardo da Vinci's Views on Atherosclerosis," presented at the Twenty-Third International Congress of the History of Medicine, London, September 2–9, 1972.

3. Few of the details of da Vinci's childhood were directly recorded, and so most, including exactly where he was born and when he moved, are subject to debate.

4. It is likely that da Vinci's father paid Verrocchio a modest sum to train da Vinci, but Verrocchio would have had to house and feed da Vinci, both of which were possible only because of the demand by the wealthy for great and ambitious art.

5. Da Vinci was aware he was doing a better job than anyone else; for instance, he described Michelangelo's nudes as "nutcracker men" and begged other artists not to copy Michelangelo's poor understanding of the body.

6. At least in the context of a book about the heart.

7. M. Kemp, "Dissection and Divinity in Leonardo's Late Anatomies," *Journal of the Warburg and Courtauld Institutes* 35 (1972): 200–25. See also Keele's beautiful treatise on da Vinci titled *Leonardo da Vinci on the Movement of the Heart and Blood* (London: Harvey and Blythe, 1952).

8. Francis Wells, "The Renaissance Heart," in J. Peto, ed., *The Heart* (London: Wellcome Collection, 2007).

9. It was a sentiment that would cause tensions with the Church. The Church allowed dissections but based on the premise that the body was just a vessel for the Holy Spirit. The vessel was all right to study, but inasmuch as da Vinci thought the actual forces of the body could be understood, his work was at odds with the Church's teaching, sacrilegious even.

10. Keele, "Leonardo da Vinci's Views on Atherosclerosis."

11. B. J. Bellhouse and F. H. Bellhouse, "Mechanisms of Closure of the Aortic Valve," *Nature* 217 (1968): 86–87.

12. Da Vinci's science stemmed from his art, but his art was also constantly influenced by his science. In his deluge drawings of the flooded river Arno, da Vinci drew eddies in the river, and I have the sense that he drew them with both the river and the heart in mind.

13. Artists, including da Vinci's mentor, had been dissecting human bodies across southern Europe at this time, but they were focused on the muscles and bones. They did not typically even approach the organs with their knives, much less consider their functions.

14. For a nice summary of our modern perspective on da Vinci's contributions to the heart, see M. M. Shoja et al., "Leonardo da Vinci's Studies of the Heart," *International Journal of Cardiology* 167 (2013): 1126–38. Among the hypotheses advanced in this work is the idea that da Vinci was prevented from discovering the circulation of the blood not by some intellectual barrier but instead because he was a bit flaky and incredibly busy. The authors reference Stefan Klein, who wrote, in *Leonardo's Legacy*, "Leonardo was pursuing so many interests that he could rarely take full advantage of his chance to solve any particular problem; he simply lacked the time to do so. In cases where an additional experiment would have given him more precise information, he was already moving on to the next unknown territory. And because he was working for himself, rather than for others, he did not devote much time to the issue of publishing his findings."

15. See K. D. Keele, "Leonardo da Vinci's Influence on Renaissance Anatomy," *Medical History* 8 (1964): 360–70.

16. It is possible that in the Mona Lisa's face, there lurks something else germane to this book—a genetic defect. The Mona Lisa's eyes are yellowed, unusually so, even among depictions of the time. It has been argued that the model for the Mona Lisa had such yellow eyes because she suffered from familial hypercholesterolemia, a genetic disorder in which cholesterol levels are many times higher than in most humans. If so, this was just one more truth da Vinci's paintbrush recorded, a truth he recorded without even being aware of it.

4. Blood's Orbit

1. J. Sawday, *The Body Emblazoned: Dissection and the Human Body in Renaissance Culture* (New York: Routledge, 1995). See also K. Park's gruesomely fascinating article "The Criminal and the Saintly Body: Autopsy and Dissection in Renaissance Italy," *Renaissance Quarterly* 47 (1994): 1–33.

2. Vesalius was very avid. Hallam says of Vesalius and his friends, "They prowled by night in charnel-houses, they dug up the dead from the graves, they climbed the gibbet, in fear and silence to steal the mouldering carcase of the murderer; the risk of ignominious punishment, and the secret stings of superstitious remorse, exalting, no doubt, in the delight of these useful but not very enviable pursuits."

3. It was to be one of the three great books published between 1543 and 1546 by scholars trained in medicine in Padua. The other two were Nicolaus Copernicus's *De Revolutionibus Orbium Coelestium*, in which the author argued the earth circled the sun, and Hieronymus Fracastorius's *De Contagione et Contagiosis Morbis*, one of the first substantive works on pathology.

4. A. Castiglioni, "Three Pathfinders of Science in the Renaissance," *Bulletin of the Medical Library Association* 31 (1943): 301–7.

5. This includes his work on anatomy, which survived thanks to just three copies, one of which Harvey appears to have seen.

6. A botanist/anatomist from Pisa, Cesalpino (1524–1603), had already done the reverse of this experiment, pinching veins and observing that when he did so, the vein below (away from the heart) rather than above the pinch swelled, suggesting that the blood in veins was moving toward the heart.

7. In truth, the average body contains 5.2 liters of blood, five big soda bottles' worth. Every minute, nearly all of this blood passes through the heart, and this rate (5 liters/minute) goes up sixfold during exercise, to about 30 liters/minute.

8. This is part but not all that moves, thanks to the heart, through blood. Blood is composed of plasma (which is mostly water), red blood cells (which carry oxygen and carbon dioxide), and a variety of necessary substances and features. Hormones travel through the blood, bearing

messages from one part of the body to another. Heat moves through the body via blood, which is warmer than the rest of the body. Nutrients also move in the blood, whether sugars, fats, vitamins, or minerals, as do wastes such as urea and valuable proteins that aid in the immune response. The blood is the body's road of many uses.

9. Just when the first life turned up is actively debated. For the most recent salvo in the debate, see S. Moorbath, "Paleobiology: Dating the Earliest Life," *Nature* 434 (2005): 155.

10. R. E. Blankenship, "Early Evolution of Photosynthesis," *Plant Physiology* 154 (2010): 434–38.

5. Seeing the Thing That Eats the Heart

1. Which, as often as not, is atherosclerosis.

2. In 1879, three French physiologists had done an experiment in which they inserted a catheter through the jugular vein of a horse, up through the vein, and into the horse's enormous heart. At its tip, the catheter had a balloon that could be inflated. The pressure on the balloon due to the horse's contracting ventricle was then used to test whether or not the ventricle contracted actively, which it did. In and of itself, this story is amazing, not least because it means that in 1879, people still did not understand the rough dynamics of the heart's pump. One of these physicians, Bernard, published a book on it, *Leçons de Physiologie Opératoire*. It was in this book that Forssmann saw the drawing.

3. This detail comes from Renee and Don Martin's *The Risk Takers*. Like much about Forssmann, it is a bit hard to confirm. Forssmann seemed uncomfortable in his skin and as a consequence told different versions of stories on different occasions. The most common version of his story derives from his published paper, and in later years, he claimed that this version (which made him look less reckless than he really was) had been a lie. I suspect that Forssmann was telling the truth later on and was lying in his youth, though it is possible it is the other way around.

4. Coronary artery disease was first described in 1768 by William Heberden, an English physician who, upon doing an autopsy on his good friend John Fothergill, another physician, observed his diseased coronary arteries and linked them to the chest pain, angina, Fothergill had felt just before his death of what we now call a heart attack. But it wasn't until Forssmann's generation that people began to realize just how common the disease was.

5. Two decades later, Forssmann did try his experiment on rabbits and when he did, it killed them. They went into immediate cardiac arrest. Had he begun with rabbits, the lesson he would have learned was that heart catheterization was always dangerous, deadly even.

6. W. Forssmann, *Experiments on Myself: Memoirs of a Surgeon in Germany* (New York: St. Martin's Press, 1974).

7. H. C. Orrin published a breathtaking book titled *The X-ray Atlas of the Systemic Arteries of the Body*. In it, he showed the blood vessels through x-ray photos. His was an update to Vesalius's drawings of the body, but one based purely on observations. *Here it is, revealed*, the images announced. One could do nothing but look on with awe.

8. M. C. Truss, C. G. Stief, and U. Jonas, "Werner Forssmann: Surgeon, Urologist and Nobel Prize Winner," *World Journal of Urology* 17 (1999): 184–86.

9. Yes, he really said this. He is actually the one who put it in print.

10. Quotes come from Forssmann, *Experiments on Myself*.

11. Roentgen discovered x-rays in 1895, but for the first years, it proved more useful as a tool of discovery in the study of cadavers than as a diagnostic tool in hospitals. With x-rays, the seemingly invisible could be made visible, especially, as in the case of cadavers, if the "patient" could hold still for the many minutes the early x-rays took. In 1920 a book by Orrin, *The X-ray Atlas of the Systemic Arteries of the Body*, was published in England, showing x-rays of cadavers who'd had dye injected into their blood vessels. The resulting imagines were arguably the first significant progress in depicting the arteries and veins since the work of Harvey and Malpighi in the late 1600s. Suddenly, the secret passages of the body were revealed, layered on top of one another. The book, like many advances in our understanding of the cardiovascular system, was as much art as science.

12. What exactly happened in the x-ray room has been retold in many different ways. Certainly the x-ray was taken. Romeis was present and angry. Forssmann might have kicked Romeis; he might not have. The x-ray technician might have objected; she might not have. Forssmann might have let out a barbaric yowl when he saw the catheter in his heart, a groaning yelp of accomplishment; he might not have. Whatever happened, the catheter made it to his heart, and a picture was snapped.

13. In this paper, Forssmann would fabricate a series of stories about the procedure in order to make it seem less radical than it really was.

14. Somewhat ironically, beginning in the 1930s, the first efforts to build on Forssmann's successes used his method to study cardiac output in human patients, just as had been done in the study that led to Forssmann's cherished image of the catheterized horse.

15. Forssmann would work to save Nazi soldiers and then, when he began to realize the full horrors of the war, those threatened by the same soldiers. Forssmann witnessed and, by his own account in his autobiography, tried to stop the shooting of six hundred Russian peasants on Whitsunday in 1942.

16. Forssmann's Nobel Prize acceptance speech can be read on the Nobel Prize website.

17. Interestingly, Forssmann had an even more ignored antecedent. In 1831, Johann Dieffenbach put a catheter into the heart of a patient dying of cholera to drain away excess blood. Or so he says. Without producing an x-ray, Dieffenbach receives little credit for being first. It was also difficult for him to inspire his own colleagues to believe that such an endeavor was possible or reasonable.

18. Sones seems to have been all these things from an early age. He studied medicine at the University of Maryland, where his professor informed him cardiology was "a nothing specialty" in which there would "never be any great discoveries." This convinced Sones to study cardiology and to make great discoveries.

19. According to David Monagan's *Journey into the Heart: A Tale of Pioneering Doctors and Their Race to Transform Cardiovascular Medicine* (New York: Gotham Books, 2007), Sones was sometimes so eager to dictate case reports that he would kick in the ladies' room door and yell to the secretaries (who might or might not have been inside), "Type, type, type."

20. For more on the personality of Sones, see Monagan, *Journey into the Heart*.

21. From a chapter on Sones, "The Way to a Human's Heart," in D. Robinson, *The Miracle Finders: The Stories Behind the Most Important Breakthroughs of Modern Medicine* (New York: David McKay, 1976).

22. By all accounts, Sones's disregard for rules and social norms was equaled only by his regard for rigorous work and for doing that work oneself. Sones started his career at Cleveland Clinic in a small, third-floor office but before long his territory had greatly expanded, in part because he worked when everyone else was sleeping, pushed out of the way anyone who was in the way, and, by force of will, did things right. Young doctors in training often admired Sones but seldom could keep up with him. See, for example, W. C. Sheldon, "F. Mason Sones Jr.—Stormy Petrel of Cardiology," *Clinical Cardiology* 17 (1994): 405–7. Sheldon also reports on Dr. William Proudfit's summary of the rules by which Sones lived, which were:

Be honest
Nothing is good enough
Find an expert
Don't read (or write)—if you must write don't use semicolons
Don't calculate
Don't rely on gadgets
Don't watch the clock
Don't repeat experiments indefinitely
Concentrate on the problem
Simplify the problem
Make a decision
Communicate

23. F. M. Sones, "Cine Coronary Arteriography," *Modern Concepts in Cardiovascular Disease* 31 (1962): 735–38.
24. G. A. Lindeboom, "The Story of a Blood Transfusion to a Pope," *Journal of the History of Medicine and Allied Sciences* 9 (1954): 455–59.
25. There had been attempts at more ambitious surgeries. In 1925, London surgeon Henry Souter attempted a valve repair. He cut through the pericardium, through the heart, and into the atria. Once in, he put his finger into the patient's inflamed mitral valve (and through into the left ventricle). As he began to sew up the valve, the heart started to flutter. It spat blood. Everything went wrong. In a panic, somehow Souter sewed the heart back up (without fixing anything), and the girl lived, but this story chastened many of those who imagined undertaking ambitious surgery.
26. There has always been someone willing to oppose progress on the heart, so it is perhaps understandable that voices of reason were ignored.

6. The Rhythm Method

1. The full quote, which is worth a read for its unusual earnestness in a scientific paper, can be found in J. H. Gibbon, "The Development of the Heart-Lung Apparatus," *Review of Surgery* 27 (1979): 231–44.
2. Gibbon appears to have been liked by nearly everyone. Rudolph Camishon said of him, "This gem of a man glistens most radiantly." He was kind and eager to do good in the world. As David Cooper wrote, he was "the tabloid press's nightmare, the paparazzi's despair, for it almost impossible to find anyone who has any criticism of him. To a large extent, he seems faultless and unblemished." D. K. C. Cooper, *Open Heart: The Radical Surgeons Who Revolutionized Medicine* (New York: Kaplan Publishing, 2010).
3. John's great-great-grandfather was John Hannum Gibbons (with an *s*). Gibbons was born in Chester County, Pennsylvania, but educated in Edinburgh, Scotland, the site of the most notorious body snatching in history in later years, all in the name of anatomy. In addition to his direct line of descent from doctors, Gibbon had physicians scattered around his family more generally: uncles, great-uncles, and the like. One of John Gibbon's nephews is now a doctor. It is what the Gibbons did and do. It seems unlikely he ever had a chance of convincing his parents to let him write, though it seems as though he had a great deal to write about.
4. Many details of Gibbon's life derive from the National Academy biography found here: http://www.nasonline.org/publications/biographical-memoirs/memoir-pdfs/gibbon-john.pdf.
5. J. H. Gibbon, "The Maintenance of Life During Experimental Occlusion of the Pulmonary Artery Followed by Survival," *Surgery, Gynecology and Obstetrics* 69 (1939): 602.

6. F. D. A. Moore, *A Miracle and a Privilege: Recounting a Half Century of Surgical Advance* (Washington, DC: John Henry Press, 1995). At about this time, a group of scientists that included Francis Moore visited the lab. Moore described a scene that was probably similar to how Jack and Maly had spent much of the last ten years of their professional life: "We were ushered into the operating room.... The pump oxygenator was approximately the size of a grand piano. A small cat, asleep on one side, was the object of all this attention. The cat was connected to the machine by two transparent blood-filled plastic tubes. The contrast in size between the small cat and the huge machine aroused considerable amusement among the audience...we began to sense that we were not walking on a dry floor. We looked down. We were standing in an inch of blood. 'Oh, I'm sorry,' said Gibbon, 'the confounded thing has sprung a leak again.'"

7. For more on Alibritten, a hero in his own right, see K. D. Hedlund, "A Tribute to Frank F. Alibritten Jr.: Origin of the Left Ventricular Vent during the Early Years of Open-Heart Surgery with the Gibbon Heart-Lung Machine," *Texas Heart Institute Journal* 28 (2001): 292–96.

8. Gibbon met Watson because he was the father-in-law of one of his colleagues.

9. Equivalent to about $255,581 in 2013, accounting for inflation.

10. Replicating a human lung is not trivial. The capillaries in the lungs absorb oxygen (and release carbon dioxide) over a surface area of six hundred square feet (above the size of a tennis court) but do so in a volume no larger than a small loaf of bread. More generally, the body has layer upon layer of maximized internal surface area. The surface area of the lungs is immense, thanks to branching bronchi. The surface area of the capillaries is immense, thanks to their own branching. But even inside cells, surfaces are expansive. Mitochondria, those creatures within our cells, are built of folded membranes that allow each individual mitochondrion to have a surface area manyfold greater than would be the case for a sphere of the same diameter, in order to better allow them to burn oxygen. A similar convolutedness is found in other organs, including the kidneys, liver, and intestines, where having more surface area is beneficial.

11. This is much simpler than the movement of oxygenated blood in the body. In the body, two large pulmonary veins leave each lung and travel to the left atrium, where the blood is pumped to the left ventricle and then out to the body.

12. According to D. K. C. Cooper, and as a measure of how much medicine has changed, the room to which Cecelia was sent to recover was a forty-bed ward, twenty beds to a side in a long room like that in which Florence Nightingale did her work. A nurse sat at a desk nearest to the sickest patient, in this case, Cecelia, who lay, recovering, beside the nurse's desk and beneath the nurse's ever-watchful gaze.

13. See a more complete description of the public response to the device in James Le Fanu, *The Rise and Fall of Modern Medicine* (New York: Carroll and Graf, 1999).

14. "Historic Operation," *Time* 61 (1953).

15. The lungs need almost as much oxygen, and hence blood, as the heart and brain, and they suffer a similar problem as the heart because the blood that arrives at the lungs is depleted of oxygen. The lungs are able to deal with this problem by having their own set of arteries, the bronchial arteries, which, like the coronary arteries of the heart, supply a high dose of oxygen-enriched blood. Like much about the cardiovascular system, these arteries were first described by da Vinci, in beautiful drawings in which they cascade over the lungs like tangled hair.

16. Others followed in the next decades with their own heart-lung machines, each a derivation of Gibbon's design. By the 1960s, heart-lung machines were standard enough that several companies had begun to produce them.

17. For the general story of the body's amazing electricity, see F. Ashcroft, *The Spark of Life: Electricity in the Human Body* (New York: W. W. Norton, 2012).

18. The link with coffee is actually still subject to debate; generally, both doctors and patients perceive a link between coffee consumption and minor heart arrhythmias. Yet the biggest studies to date on such a link have found nothing.

19. The possibility of clots is why patients with atrial fibrillation are given blood thinners such as warfarin.

20. This is what was first tried, unsuccessfully, on my mom.

21. It is because of ventricular fibrillation that paddles are most often used on the heart. The paddles are used not to restart the heart but instead to briefly stop it, in the hopes that when it restarts naturally, it will do so with a more normal rhythm.

22. Heart block is another problem of the heart's electricity, but one that historically was very rare. Heart block occurs when the signal between the atria and ventricles is impaired (the ventricular node does not get the signals it is supposed to). Heart block became more common as a result of open-heart surgeries, which can cause interference with electrical signals. Heart block is not, in and of itself, necessarily fatal. Enough blood falls from the atria to the ventricles to allow the heart and its function to continue, but the heart and the patient's life slow down.

23. H. G. Mond and A. Proclemer, "The Eleventh World Survey of Cardiac Pacing and Implantable Cardioverter-Defibrillators: Calendar Year 2009—a World Society of Arrhythmia's Project," *Pacing and Clinical Electrophysiology* 34 (2011): 1013–27.

24. The Chimera has even become a kind of unofficial symbol of organ transplantation, the ugly amalgamated beast of the past a symbol of the hoped-for

future; see R. Kuss and P. Bourget, *An Illustrated History of Organ Transplantation* (Rueil-Malmaison, France: Laboratoires Sandoz, 1992).

25. J. Dewhurst, "Cosmas and Damian, Patron Saints of Doctors," *Lancet* 2 (1988): 1479.

26. A. Carrel and C. Lindbergh (yes, the flying one), "Culture of Whole Organs," *Science* 31 (1935): 621.

27. A Kansas surgeon, J. R. Brinkley, single-handedly performed more than 16,000 such transplants, mostly with gonads from goats. Brinkley was eventually barred from practicing medicine, after which he lost a close race for governor; see D. Hamilton, *The Monkey Gland Affair* (London: Chatto and Windus, 1986). See also F. Lydston, "Sex Gland Implantation: Additional Cases and Conclusions to Date," *Journal of the American Medical Association* 94 (1930): 1912.

7. Frankenstein's Monsters

1. Interestingly, it remains unclear why a dog's heart is less supple than a human's. This is not the only difference between dogs and humans in terms of cardiovascular systems either; for example, dog lungs are much more prone to collapse during surgery.

2. "Surgeons Repair Hearts of Four Dogs," *New York Times*, April 19, 1962.

3. Mary writes of this influence: "Many and long were the conversations between Lord Byron and Shelley, to which I was a devout but nearly silent listener. During one of these, various philosophical doctrines were discussed, and among others the nature of the principle of life, and whether there was any probability of its ever being discovered and communicated. They talked of the experiments of Dr. Darwin [Erasmus] who reserved a piece of vermicelli in a glass case, till by some extraordinary means it began to move with voluntary motion. Not thus, after all, would life be given. Perhaps a corpse would be re-animated; galvanism had given token of such things: perhaps the component parts of a creature might be manufactured, brought together, and endured with vital warmth." Yes, you read that correctly: Mary Shelley was inspired by Erasmus Darwin, who was inspired by a rotting noodle.

4. Here, the linguistic story of the emotional heart is as rich as the biological one. One can be, if contented, lighthearted, bright-hearted, happy-hearted. One can take heart or be heartened. One can have a peaceful, quiet, or restful heart. One's heart can be at ease. It can be at rest. It can be healthy. Something can be heart-cheering, it can do the heart good or make it leap or dance for joy. The heart can swell or even burst; it can also rejoice. In passion, one can be a sweetheart or a heartthrob. Meanwhile, if things go wrong, one can be heavyhearted, downhearted, disheartened,

have a heart of darkness, a stricken heart, a struck heart, a broken heart, an achy heart, or a wrenched heart. Something can eat at the heart, and, if it does, rend it, cause it to sink, even to fall into one's boots.

5. M. C. Truss, C. G. Stief, and U. Jonas, "Werner Forssmann: Surgeon, Urologist, and Nobel Prize Winner," *World Journal of Urology* 17 (1999): 184–86.

6. At this time in the United States, most hospitals still required that donors were dead, even with regard to their hearts, before their organs could be transplanted. This meant that Shumway, Lower, and other surgeons had to wait for hearts to stop before transplanting them. This both lowered the odds of getting a heart and made it less likely that the heart transplant would succeed.

7. Initially, the care of living heart cadavers was very simple, but with time it would become clear that these "patients" required all of the care given to normal patients and more. Because their brains could not control their blood pressure, it had to be attended to, as did their hormone levels.

8. Although Shumway did not, at the time, publicly express his disappointment about what had happened, others did so on his behalf. James Hardy, the man who transplanted the heart of a chimpanzee into the body of a man, an odd voice of moderation, wrote, "My disappointment is enormous, though not so much for myself personally. I know that Norman Shumway's group at Stanford have done the most extensive and the best work in this field. We have long been waiting for them to transplant a heart from one man to another, following which, after more considered research, my team hoped to emulate them. We were technically ready long before Barnard, but we were burdened by the need to protect our public from the possible failure of such a great experiment. Shumway did everything by the book—only to have history stolen from him."

9. Kantrowitz was, during this period, asked to leave the small Jewish hospital, Maimonides, where he worked. This was clearly a difficult time for the man, but years later, in reflecting on his role in the story of heart transplants, he was less humble. "Galileo," Kantrowitz said, "was fired; they really gave him hell when he said that maybe the Earth is not the center of the university." Kantrowitz went on to compare heart transplants to Galileo's heliocentrism, suggesting that transplants too would, although initially seen as radical, eventually succeed.

10. After his retirement, Shumway, in reflecting upon Barnard, indicated to Donald McRae in his book *Every Second Counts* that Barnard was haunted by what he had done in turning heart transplants into a spectacle and stealing from Shumway and Lower what all seemed to agree they had earned. Shumway, however, in that same interview said he had come to believe that "maybe it was a blessing that we weren't first...we had enough

trouble anyway dealing with the press and all that hoohah. Boy, we had plenty of trouble. So maybe, in the end, it all worked out for the best."

11. The longest lived of these twenty-three living recipients in 1970 was one of Lower's patients, Louis B. Russell Jr., a teacher who would go on to have six additional good years of, to paraphrase Russell, living hard, eating well, and making love.

12. Wilder's story is fascinating. He was the grandson of slaves and would go on to many forms of greatness, including a term as Virginia's governor, but also most recently two terms as the mayor of Richmond, Virginia. He also founded the U.S. National Slavery Museum. Today he is an adjunct professor of public policy at Virginia Commonwealth University.

13. Of course, this was the Western definition of life. Life and death are defined in many ways in many places. To his horror, the anthropologist Colin Turnbull was famously declared dead by the Pygmies with whom he was living. The good news was that the Pygmies believed there were seven different kinds of death, and since Turnbull was in one of those stages from which it was possible to recover, they did not bury him.

14. Wada was eventually found to have "lied to the media" and "tampered with the valves in the recipient's original heart to exaggerate their defectiveness." No new heart transplants would be permitted in Japan until 1999.

15. R. Converse, "But When Did He Die: *Tucker v. Lower* and the Brain-Death Concept," *San Diego Law Review* 424 (1974–1975): 424–35. Because this case was in trial court, it does not serve as a formal legal precedent, but culturally, it set a precedent, one that has been revisited each time the issue of defining death has reemerged, as one suspects it will continue to do forever.

16. At the time, many doctors still viewed either the death of the brain or the death of the heart as a sufficient marker of the death of an individual. In this light, Lower, Barnard, and others were literally bringing their patients back to life, at least for a few days.

17. The State of Virginia voted in legislation to formalize brain death as a medical and legal concept, as, ultimately, did many other states. Another major moment in the debate about brain death came in California: Andrew Lyons shot Samuel Mitchell Allen in the head. Allen was taken to the hospital where, upon his being pronounced brain-dead, his heart was harvested for transplant into Blain Wixom by Shumway. The murderer then contended that he had not murdered the other man because his heart was still alive. The jury convicted Lyons, which led to a California law defining *death* specifically as brain death.

18. "Heart Transplant Decision Questioned," *Lakeland Ledger,* June 5, 1972.

19. In his book *Invasion of the Body*, the surgeon Nicholas L. Tilney mounts a defense of the actions of those who raced to transplant hearts. I'll let you choose how to weigh it. "A central theme," said Tilney, "of these often

troubling events is of surgeons desperately trying all means possible to salvage a handful of dying patients using concepts and techniques still in their infancy. At such climactic moments, the individual responsible must be totally convinced of himself and his talents to make instantaneous and sometimes irrevocable decisions, cast aside philosophical, religious, or societal considerations, and leave debate about correctness, appropriateness, and even the ethics of the decision to the future."

20. They dominated the field until Lower bought a cattle ranch in Montana, where he spent his retired years taking care of three hundred cattle, by himself. The animals' big hearts needed no help. He fed them, moved them, and cured them of their parasites, but he never again did surgery.

21. We think of rejection as bad, but it is really just the immune system doing part of what it has evolved to do: recognizing foreign cells. The problem with rejection is that heart transplants, however ingenious, are at odds with the evolutionary history of the body, one in which a foreign cell was always seen as dangerous. The presence of a transplanted organ triggers what is, in essence, an attack against foreign cells, the severity of which depends in part on how different (in each of many ways) the cells of the donor are from the cells of the recipient. The first step in reducing rejection was to choose a donor whose cells were as similar as possible to those of the potential recipient.

22. H. Schwartz, "A Long Shot, and Still Running: Heart Transplants," *Lakeland Ledger*, August 26, 1973.

23. J. F. Borel, "The History of Cyclosporin A and Its Significance," in D. J. G. White et al., eds., *Proceedings of an International Symposium on Cyclosporin A* (Amsterdam: Elsevier, 1972).

24. See, for example, "European Multicentre Trial, Cyclosporin in Cadaveric Renal Transplantation: One Year Follow-Up of a Multicentre Trial," *Lancet* 2 (1983): 986.

25. For Hodge's version of the story, see K. T. Hodge, S. B. Krasnoff, and R. A. Humber, "*Tolypocladium inflatum* Is the Anamorph of *Cordyceps subsessilis*," *Mycologia* 88 (1996): 715–19.

26. The asexual, soil-dwelling stage of this fungus goes by another name, *Tolypocladium inflatum*. Fungi are complex, and so, it seems, are the folks who make a living naming them.

8. Atomic Cows

1. For a nuanced depiction of the personalities of DeBakey and Cooley and their interactions, see D. K. C. Cooper, *Open Heart: The Radical Surgeons Who Revolutionized Medicine* (New York: Kaplan Publishing, 2010). Although the men would eventually reconcile, for most of their professional lives,

they were the two "big men" competing in Texas medicine, each with an outsize personality befitting the outsize state.

2. The broader history of prostheses is fascinating and worth a read; see A. J. Thurston, "Paré and Prosthetics: The Early History of Artificial Limbs," *ANZ Journal of Surgery* 77 (2007): 1114–19.

3. This is roughly the number of people on heart-transplant waiting lists in the United States in an ordinary year.

4. The National Heart Institute was created in 1948 during Truman's administration to research the causes, prevention, diagnosis, and treatment of heart and other cardiovascular diseases (including strokes). Today, the NHI lives on as the National Heart, Lung, and Blood Institute (NHLBI).

5. When a nucleus of plutonium 239 is hit by a neutron, it undergoes fission (releasing an enormous quantity of energy). It also releases additional neutrons, which hit adjacent atoms and lead to a chain reaction that runs out of control.

6. The director of the Artificial Heart Program had asked for $2 million in the first year, $8 million in the second year, and $100 million in each subsequent year, all in 1965 dollars.

7. According to Shelley McKellar, a historian who has written about this episode, "The agencies could not agree on management jurisdiction or the approach for engine development, making a collaborative venture practically impossible." Also, they seemed to hate each other.

8. The other use envisioned for plutonium 238 was in generating long-term electricity in space.

9. The first sixteen patients Kolff tried the machine on died. The seventeenth was a Nazi collaborator, Sophia Schafstadt; she lived, but in the weeks after the war, many of those who worked with Kolff wished she had not. Kolff himself had supported the Dutch resistance during the war.

10. Plutonium 238 had the advantage of a long half-life and relatively low cost. Interestingly, the cost of plutonium 238 was tied to the number of nuclear reactors in the U.S. The more nuclear the U.S. went in terms of energy, the cheaper atomic hearts would become.

11. On April 27, 1970, one human subject in Russia actually received a nuclear pacemaker inside which 165 milligrams of plutonium were enclosed and shielded. Nuclear pacemakers eventually came to the U.S. as well. Fifteen were implanted in just two days. By 1979, nearly three thousand had been implanted worldwide.

12. See discussion of this whole episode in N. L. Tilney, *Invasion of the Body* (Cambridge, MA: Harvard University Press, 2011).

13. Quoted in R. C. Fox and J. P. Swazey, *The Courage to Fail* (Chicago: University of Chicago Press, 1974).

14. The mechanical heart itself is now stored in a cabinet at the National Museum of American History, where, according to Alex Madrigal, one

can still see, on the tubes leading out of the device's twin pumps, a little of Karp's blood. For the official report, see W. C. DeVries et al., "Clinical Use of the Total Artificial Heart," *New England Journal of Medicine* 310 (1984): 273.

15. The National Heart Institute and Baylor both held inquiries into Cooley's actions. Liotta was suspended. Cooley voluntarily left his university position, though he continued to be among the world's most active heart surgeons and innovators. In fact, by losing his position, he was actually freed up to make even more money than he had while at Baylor.

16. W. J. Kolff and D. B. Olsen, "Testing of Radioisotope-Powered Mechanical Heart in Calves," *Biomedical Engineering Support Progress Report*, August 15, 1976–May 15, 1977.

17. The plutonium in artificial hearts was dangerous for several reasons. There was the (very low) chance it would leak in the body. But there were also two more obvious problems. Once someone with an atomic heart died, what happened to the plutonium? Old plutonium is still radioactive; it is also toxic, one of the most poisonous substances on Earth.

18. "The Glamorous Artificial Heart," *New York Times*, January 15, 1983.

19. In one recent study of 133 people given assist devices, the average time they stayed on those devices was 180 days, and 100 of the 133 patients eventually received heart transplants. Twenty-five of the remaining 33 died, but the hearts of a few individuals recovered enough for them to live without the support device. See L. W. Miller et al., "Use of a Continuous-Flow Device in Patients Awaiting Heart Transplantation," *New England Journal of Medicine* 357 (2007): 885.

9. Lighter than a Feather

1. http://www.yare.org/essays/The%20Tomb%20of%20Queen%20Meryet .htm.

2. A. R. David, A. Kershaw, and A. Heagerty, "Atherosclerosis and Diet in Ancient Egypt," *Lancet* 5 (2010): 718–19.

3. Among those likenesses that have been preserved is one that speaks to the heart's link to emotion. An obelisk dedicated to Hatshepsut Karnak reads, "Now my heart turns this way and that, as I think what the people will say. Those who see my monuments in years to come, and who shall speak of what I have done."

4. H. E. Winlock, *The Tomb of Queen Meryet-Amun at Thebes* (New York: Metropolitan Museum of Art, 1932).

5. The heart pounds with excitement because of exciting events, whether they be things that have happened (you were chased by a tiger) or things that might happen (you think you hear a tiger, and it seems prudent to be prepared). As

part of its fight-or-flight response, the body releases endorphins, which speed up the heart in order to pump more blood and oxygen to the cells that will need them in the event that you need to (a) fight a tiger or (b) run.

6. Winlock did not know it, but his original quarry, Queen Hatshepsut, had been found decades before, on the floor of a minor tomb in the Valley of the Kings alongside another woman and mummified geese, but she had never been studied because her body was so unadorned and unremarkable. Her son, who begrudged her her success, had had her body moved to this locale, hoping in this and other ways to erase her traces. He nearly did so, but one of her teeth had been collected when her tomb was first found, and the DNA in that tooth, when studied in 2005, revealed her true identity.

7. Or was heavier than the goddess Maat, who was charged with assessing lawfulness, morality, good, and evil.

8. W. B. Ober, "Weighing the Heart Against the Feather of Truth," *Bulletin of the New York Academy of Medicine* 59 (1979): 636–51.

9. Heart disease due to atherosclerosis is often called ischemic heart disease, where *ischemic* derives from the Greek word for "narrowing" or "restriction."

10. The name derives from the history of cholesterol's discovery. It was originally discovered inside gallstones and so associated with bile.

11. There was some earlier work, particularly that of Marc Ruffer, who found atherosclerosis in some mummies in the early 1900s, but comparing Ruffer's older studies to our modern definitions of the symptoms of atherosclerosis was not simple. See M. A. Ruffer, "On Arterial Lesions Found in Egyptian Mummies," *Journal of Pathological Bacteriology* 16 (1911): 453–62.

12. See figure 2 in A. H. Allam et al., "CT Studies of the Cardiovascular System in Ancient Egyptian Mummies," *American Heart Hospital Journal* 10 (2010): 10–13.

13. A. H. Allam et al., "Atherosclerosis in Ancient Egyptian Mummies: The Horus Study," *Journal of the American College of Cardiology: Cardiovascular Imaging* 4 (2011): 315–27.

14. W. A. Murphy et al., "The Iceman: Discovery and Imaging," *Radiology* 226 (2003): 614–29.

15. R. C. Thompson et al., "Atherosclerosis Across 4,000 Years of Human History: The Horus Study of Four Ancient Populations," *Lancet* 381 (2013): 1211–22.

10. Mending the Broken Heart

1. Favaloro once said in an interview (D. K. C. Cooper, *Open Heart: The Radical Surgeons Who Revolutionized Medicine* [New York: Kaplan Publishing,

2010]) that when he was in college, his professor told him that to be a great surgeon, one needed to become a great carpenter. Like many of the stories doctors tell about themselves, this one seems to mark Favaloro early as fated for the life he led, a life as a carpenter of the heart.

2. Often, the first symptom of a permanently blocked coronary artery is a heart attack. Often, the second is death.

3. In practice, it tended to leave patients with the original angina and, as a bonus, a complete and permanent lack of energy.

4. See the description in K. L. Greason et al., "Myocardial Revascularization by Coronary Arterial Bypass Graft: Past, Present, and Future," *Current Problems in Cardiology* 36 (2011): 325–68.

5. D. J. Fergusson et al., "Left Internal Mammary Artery Implant— Postoperative Assessment," *Circulation* 37 (1968): 24–26.

6. G. Murray et al., "Anastomosis of a Systemic Artery to the Coronary," *Canadian Medical Association Journal* 71 (1954): 594–97.

7. On May 2, 1960, Robert Goetz actually performed bypass surgery on a thirty-eight-year-old man. A coronary artery was replaced by a mammary artery that was held in position by an artificial ring. The surgery appears to have been successful, but the patient died a year later due to a heart attack and there was no autopsy to see how or whether the death was related to the original surgery. It is unclear if Favaloro knew of this surgery.

8. H. E. Garrett, E. W. Dennis, and M. E. DeBakey, "Aortocoronary Bypass with Saphenous Vein Graft: Seven-Year Follow-Up," *Journal of the American Medical Association* 223 (1973): 792–94.

9. Blockage of the left coronary artery is more serious, serious enough that the blocked vessel is often described, in the dark humor of surgeons, as a widow maker.

10. For example, mammary artery grafts now appear more successful than vein grafts and so have become the standard.

11. All three men would receive many accolades. One in particular is worth mentioning for Sones. He was awarded the Galen Medal of the ancient Worshipful Society of Apothecaries of London. Upon being told of the award, he was asked if he knew who Galen was. He replied, "Oh yeah, Galen, I remember him. He was number sixty-two in my class in medical school, and I was number sixty-one."

12. Now referred to, oh so mundanely, as "plain old balloon angioplasty" (POBA).

13. J. G. Motwani and E. J. Topol, "Aortocoronary Saphenous Vein Graft Disease: Pathogenesis, Predisposition, and Prevention," *Circulation* 1998 (1998): 916–31.

11. War and Fungus

1. Of note, this is not high for Endo's age. It is only high relative to a mean for all individuals of all ages and cultures.
2. In the Akita region of Japan, there is a long history of gatherers. As recently as six hundred years ago, hunter-gatherer tribes roamed the forested hills of the area.
3. This genus of fungus, *Amanita*, contains many of the most deadly mushrooms in the world, including the death cap.
4. I have to remind myself that such interest *is* unusual. It is the sort of curious interest on which my field depends, an interest my colleagues and I all share, and the sort of interest that might tempt us to repeat this experiment in the kitchen down the hall from my lab at North Carolina State University. (We did.)
5. A. Endo, "A Historical Perspective on the Discovery of Statins," *Proceedings of the Japan Academy, Series B, Physical and Biological Sciences* 86 (2010): 484.
6. It sucks being a lab rat.
7. For more on the statin story, see J. J. Lie, *Triumph of the Heart: The Story of Statins* (New York: Oxford University Press, 2009).

12. The Perfect Diet

1. For more about the high-elevation work, see S. W. Tracy, "The Physiology of Extremes: Ancel Keys and the International High Altitude Expedition of 1935," *Bulletin of the History of Medicine* 86 (2012): 627–60.
2. Ancel Keys, "Notes on the Laboratory of Physiological Hygiene, University of Minnesota," February 9, 1945, Ancel Keys Collection, University of Minnesota Archives, Minneapolis, 3.
3. For more about this astonishing experiment and all of its ethical complexities, see T. Tucker, *The Great Starvation Experiment* (New York: Free Press, 2006).
4. Heavy cholesterol (HDL—high-density lipoproteins) helps break down plaques in the blood and move light (LDL—low-density lipoproteins) cholesterol out of the way, carrying it back to the liver. LDL cholesterol, however, is the one with the predilection for getting stuck on artery walls and precipitating the reactions that lead to atherosclerosis.
5. In light of Queen Meryet-Amun's story, he was clearly wrong that high cholesterol and atherosclerosis were new. It is less clear whether or not he was right that there was a postwar boom in atherosclerosis and heart disease. The data before the war are too limited to inspire confidence, and while there was a peak in heart attacks and heart disease, part of this peak may be due to the waning of other diseases rather than the waxing of heart disease.

6. Ideally, he would have chosen people who were similar genetically and then prescribed each of those people one of a series of diets that varied in the number of calories; percentage of fats, proteins, and carbohydrates; and composition of other dietary particulars—fish oils, olive oils, nuts, vegetables, and fruits. He would need enough people to have some individuals in each treatment who would suffer from heart attacks. If no one had a heart attack, there would be nothing to compare. If Keys were to experiment with, let's say, fifteen diets, he might need as many as fifteen hundred people to be part of the experiment. He would also have to convince people to stick with an experimental diet their entire lives. What was more, the experiment would work only if those people did not have the chance to choose which diet they would consume for the rest of their lives! No one has done such an experiment. No one ever will. Therefore, all studies that link diet, lifestyle, genetic background, and health outcomes are, at best, suggestive.

7. At about this same time, Christiaan Barnard, in his book *Heart Attack*, described Keys as one of the most outstanding nutritionists and epidemiologists of modern times.

8. One of the very few diets no one seems to advocate is a diet engineered by science. In 1940, it seemed clear that technology would produce the perfect diet, supplemented with each thing it needed. This no longer appears to be held as a truth except in the military and in attempts to deal with starvation. We now seem to believe that technology cannot solve the problems of diet; we believe this even as we, each year, consume more and more technologically processed foods.

9. R. Estruch et al., "Primary Prevention of Cardiovascular Disease with a Mediterranean Diet," *New England Journal of Medicine* 368 (2013): 1279–90.

10. Though moving did not solve all of their problems. As the Keyses aged, they ultimately had to leave Italy. They were not Italian, so they had no family there to care for them. A new culture is no substitute for family.

13. The Beetle and the Cigarette

1. *Cardio* comes from the Greek for "heart," *-ologist* is someone who studies. One might then imagine that the term *cardiologists* would be reserved for those who study the heart. It is not. In practice, cardiologists are physicians who mend hearts without using open-heart surgery. They are ever more common. Cardiac surgeons, by contrast, that bloody-handed and dexterous clan, are ever more rare.

2. Christiaan Barnard even derisively referred to those who used statistics to evaluate the relative benefits of different treatments as "number boys." He, like many surgeons, had much more confidence in his own intuition than in data.

3. European Coronary Surgery Study Group, "Long-Term Results of Pro- spective Randomised Study of Coronary Artery Bypass Surgery in Stable Angina Pectoris," *Lancet* 316 (1982): 1173–80.
4. In other words, patients in the study had randomly, like guinea pigs, been assigned to one treatment arm or another. The advantage of such an approach is that it removes any preexisting differences in terms of who gets one treatment and who gets another.
5. Here it is worth noting that because the effect of these pollutants is due purely to their size, it does not much matter what they are. So long as they are small, they will cause problems.
6. T. Takano, K. Nakamura, and M. Watanabe, "Urban Residential Envi- ronments and Senior Citizens' Longevity in Megacity Areas: The Impor- tance of Walkable Green Spaces," *Journal of Epidemiology and Community Health* 56 (2002): 913–18.
7. D. J. Nowak, D. E. Crane, and J. C. Stevens, "Air Pollution Removal by Urban Trees and Shrubs in the U.S.," *Urban Forestry and Urban Greening* 4 (2006): 115–23.

14. The Book of Broken Hearts

1. In 1976, still without so much as an undergraduate degree, Thomas was appointed as an instructor in surgery at Johns Hopkins.
2. It was Osler who wrote that the tragedies of life are largely arterial, and he should know; he might have specified adult life, though. The tragedies of children are of a different nature.
3. Despite these and many other successes, Abbott was never promoted beyond the role of assistant professor. In addition, she was (against her wishes) removed from all of her teaching responsibilities in 1923, before she wrote her great book. After death, she would receive more praise. The museum where she worked is now called the Maude Abbott Medical Museum.
4. W. N. Evans, "Helen Brooke Taussig and Edwards Albert Park: The Early Years (1927–1930)," *Cardiology in the Young* 20 (2010): 387.
5. As told by Taussig's friend and student Charlotte Ferencz after Taussig's death. See D. G. McNamara (who trained with Taussig and later worked in Texas with both Cooley and DeBakey) et al., "Helen Brooke Taussig: 1898 to 1986," *Journal of the American College of Cardiology* 10 (1987): 662–71.
6. L. Malloy, "Helen Brooke Taussig (1898–1986)," in J. Bart, ed., *Women Succeeding in the Sciences: Theories and Practices Across Disciplines* (West Lafayette, IN: Purdue University Press, 2000).
7. The chronically ill children were housed in the Harriet Lane Home for Invalid Children, named for and funded by Harriet Lane, the niece of President Buchanan, whose two children both died of rheumatic fever, the

disease whose successful treatment with antibiotics would result in it no longer leading to chronic illness.

8. Gross was an excellent and creative surgeon, but a complex human. By some accounts he was manic-depressive, disappearing for long weeks without explanation from both his wife and his work, and then reappearing as though nothing had happened. On the one hand, Taussig was not the only person whose career Gross failed to support. On the other hand, some (though they appear to have been few) who worked with Gross swore by both his genius and his supportiveness. For more on Gross, see D. K. C. Cooper, *Open Heart: The Radical Surgeons Who Revolutionized Medicine* (New York: Kaplan Publishing, 2010).

9. See the original paper describing the first three surgeries: A. Blalock and H. B. Taussig, "The Surgical Treatment of Malformations of the Heart in Which There Is Pulmonary Stenosis or Pulmonary Atresia," *Journal of the American Medical Association* 128 (1945): 189–202.

10. "First Blue Baby Operation Tried Two Years Ago Today," *Miami News*, November 29, 1946.

11. Letter from Lord Brock to Mark Ravitch (September 1965), cited in R. Hurt, *The History of Cardiothoracic Surgery: From Early Times* (New York: Parthenon, 1996).

12. This accomplishment is all the more important in light of its context. In the 1960s and 1970s, about two or three out of every thousand cardiothoracic surgeons in the United States were women. In the 1980s, the number began to increase, but modestly. It remains around 2 or 3 percent. See S. Roberts, A. F. Kells, and D. M. Cosgrove, "Collective Contributions of Women to Cardiothoracic Surgery: A Perspective Review," *Annals of Thoracic Surgery* 71 (2001): 19–21.

13. In addition to her work on the heart, Taussig also played a key role in calling attention to the dangers of thalidomide. Her testimony before Congress prevented its being approved in the United States.

15. The Evolution of Broken Hearts

1. L. K. Altman, "Dr. Helen Taussig, 87, Dies; Led in Blue Baby Operation," *New York Times*, May 22, 1986. For more details about Taussig's life and death, see D. G. McNamara et al., "Helen Brooke Taussig: 1898 to 1986," *Journal of the American College of Cardiology* 10 (1987): 662–71.

2. This is the same sort of challenge human societies face. When one or a few individuals live together, no roads are necessary. Everything needed can be gathered close to hand (and disposed of there as well). But as societies grow, roads and plumbing became necessary: roads to reach faraway resources and plumbing to take away the waste these resources created.

3. For more on the evolution of the heart, see Carl Zimmer's excellent "The Hidden Unity of Hearts" in the April 2000 issue of *Natural History Magazine*.

4. Colleen Farmer at the University of Utah thinks she understands this mystery of the lungfish story. She thinks lungfish succeeded and diversified for hundreds of millions of years, but then they met up with real monsters. The descendants of the lungfish that crawled onto land, terrestrial vertebrates, had evolved and diversified. By 220 million years ago, they had even begun, in the form of pterosaurs, to fly. The jaws of some pterosaurs suggest they evolved to eat fish at the surface of the water. It would have been easiest for them to catch lungfish, those fish that had to keep coming up for air. Once flying predators evolved, lungfish had fewer advantages in the sea. They could swim faster, sure, but they had to keep surfacing, and when they did, they were eaten by some of the most ferocious flying beasts ever to have lived. When they did, lungfish began to go extinct or to rely less on their lungs. Modern fish are a lineage of lungfish that lost their lungs entirely.

5. Here, I am focusing on just the big extant (living) lineages of land vertebrates. The full evolutionary tree of land vertebrates includes many stories of the heart, but most of those stories are very difficult to study because they played out inside the bodies of organisms, such as dinosaurs, that are extinct and in which the soft tissue of the heart was not preserved.

6. According to Taussig's hypothesis, because lizards, snakes, turtles, mammals, and birds all descend from a common ancestor with a heart with two atria and one ventricle, they should all share similar atrial deformities. This has not been tested in any formal way, but mammals and birds can both be born with holes in the wall (septum) between the atria. At least some lizards can too.

7. Functionally divided ventricles and associated coronary arteries also arose, independently, in another active group of vertebrates, the varanid lizards, which include komodo dragons. So next time you are tempted to tease a komodo, remember that it can run for a while because it has a four-chambered heart. Coronary arteries also appear to have arisen in some fish capable of prolonged activity, such as mackerels and tunas. At the opposite extreme, some slow-moving bottom-feeders like carp have no coronary circulation at all and go weeks or even months without any oxygen in their hearts.

8. This idea, often referred to as "ontogeny recapitulates phylogeny," was long thought to be a generality in evolution, so much so that developmental stages were used to infer our ancestry. Generally speaking, our developmental stages do not rehash our evolution. However, in the specific case of our hearts, it seems as though the heart has become more complex in part by adding steps to development.

9. M. L. Kirby, T. F. Gale, and D. E. Stewart, "Neural Crest Cells Contribute to Aorticopulmonary Septation," *Science* 220 (1983): 1059–61. Here one interesting exception is that some scientists, including Margaret Kirby, have begun to use chickens as models of congenital deformities of ancient parts of the hearts shared by birds and mammals, such as those related to the pulmonary artery.

10. In this special case, development repeats evolution's trajectory. During development, the coronary arteries begin to grow only once the heart is big enough for some cells to start to become anoxic. Anoxia is a signal that the coronary arteries are necessary in each fetal bird or mammal. In humans, the coronary arteries are fully developed about three to six weeks into gestation.

16. Sugarcoating Heart Disease

1. For example, see L. Munson and R. J. Montali, "Pathology and Diseases of Great Apes at the National Zoological Park Zoo," *Zoo Biology* 9 (1990): 99–105.

2. R. Margreiter, "Chimpanzee Heart Was Not Rejected by Human Recipient," *Texas Heart Institute Journal* 33 (2006): 412.

3. For example, see L. J. Lowenstine, "A Primer of Primate Pathology: Lesions and Nonlesions," *Toxicologic Pathology* 31 (2003): 92–102.

4. Which closed in 2012 (http://carta.anthropogeny.org/museum/collections/pfa).

5. Fortunately, both Yerkes and the foundation kept excellent records of just what had been squirreled away, when, and in what context.

6. I once cleaned out a freezer at the University of Connecticut and found, at the very back, a bag marked *S. Cunningham — Costa Rican air, do not open.* I, of course, opened it; it smelled just like Costa Rica.

7. At Yerkes, the average cholesterol level among 106 male chimps—across their lifetimes—was 211.1 mg/dl, which is regarded as borderline high in humans. Similar levels were observed at two other centers. On average, cholesterol levels are higher in chimps than in humans (even Western couch potatoes), but the biggest difference between humans and chimps in terms of cholesterol is timing. Cholesterol is high in infant chimpanzees and stays high, whereas cholesterol is low in young humans but increases over the years.

8. In this treatment, a horse is given white blood cells from a human. The horse then develops antibodies to the white blood cells. The horse serum, with those antibodies, is then injected into the patient. In theory, the antibodies in the horse's serum are supposed to placate the patient's immune system—which in aplastic anemia is overactive. In practice, no one knows why (or even really if) this treatment works.

9. H. Higashi et al., "Antigen of 'Serum Sickness' Type of Heterophile Antibodies in Human Sera: Identification as Gangliosides with N-Glycolylneuraminic Acid," *Biochemical and Biophysical Research Communications* 79 (1977): 388–95.

10. See interview in J. Cohen, *Almost Chimpanzee: Searching for What Makes Us Human* (New York: Henry Holt and Company, 2010).

11. A. Varki et al., *Essentials of Glycobiology*, 2nd edition (New York: Cold Spring Harbor Laboratory Press, 2009).

12. E. Muchmore et al., "Developmental Regulation of Sialic Acid Modifications in Rat and Human Colon," *FASEB Journal* 1 (1987): 229–35.

13. H. H. Chou et al., "A Mutation in Human CMP-Sialic Acid Hydroxylase Occurred After the Homo-Pan Divergence," *Proceedings of the National Academy of Sciences* 95 (1998): 11751–56.

14. E. A. Muchmore, S. Diaz, and A. Varki, "A Structural Difference Between the Cell Surfaces of Humans and the Great Apes," *American Journal of Physical Anthropology* 107 (1998): 187–98.

15. At the time, our differences were so little understood that the director of Yerkes, Thomas Insel, remarked in an interview in 1998 with the journal *Science*, "You could write everything we knew about the genetic differences (between humans and chimps) in a one-sentence article." See A. Gibbons, "Which of Our Genes Make Us Human?" *Science* 281 (1998): 1432–34.

16. I focus here on how eating mammal meat (and mammal sialic acid) affects inflammation and heart disease, but this is not the only effect. Some deadly strains of *E. coli* bacteria produce toxins when they infect human bodies; it is the toxins that make these bacteria dangerous. One of the toxins these bacteria produce (known as subtilase cytotoxin, or SubAB) binds to the standard mammalian sialic acid. As a result, individuals who eat mammal meat are at risk from these toxins, but those who do not are not. See J. Cohen, "Eat, Drink, and Be Wary: A Sugar's Sour Side," *Science* 31 (2008): 659–61, and P. Tangvoranuntakul et al., "Human Uptake and Incorporation of an Immunogenic Nonhuman Dietary Sialic Acid," *Proceedings of the National Academy of Sciences* 100 (2003): 12045–50.

17. D. H. Nguyen et al., "Loss of Siglec Expression on T Lymphocytes During Human Evolution," *Proceedings of the National Academy of Sciences* 103 (2006): 7765–70; P. C. Soto et al., "Relative Over-Reactivity of Human Versus Chimpanzee Lymphocytes: Implications for the Human Diseases Associated with Immune Activation," *Journal of Immunology* 184 (2010): 4185–95. A set of proteins called siglecs, which Ajit Varki discovered, bind to sialic acids. With the change in human sialic acids, these proteins have become less common in our bodies, which is relevant because the role of these compounds appears to be to put the brakes on the immune system,

calming its response and encouraging it to offer peace rather than wage war.

18. Humans host several kinds of mites, including scabies mites (which are a problem) and *Demodex* mites (which my lab studies and which tend not to be a problem). These small creatures appear to be found on all adult humans, but we are constantly being colonized by different strains. Humans lack a third type of mites found on most other apes: fur mites. All other apes are covered with them. They are one of the groups of parasites we seem to have lost when we shed our fur.

19. If the heart disease seen in chimps is due to a virus (which it seems it might be), by losing sialic acid, we may have escaped the sort of heart disease chimps face.

20. These, like all measures of life expectancy, are averages.

21. It has been argued, quite plausibly, that the decrease in life expectancy associated with agriculture is better described as being associated with a major social transition. Transitions, it is suggested, predictably decrease life expectancy, as all hell breaks loose during the shift from one way of life to another.

22. For more on the shifting ecology of teeth, see C. J. Adler et al., "Sequencing Ancient Calcified Dental Plaque Shows Changes in Oral Microbiota with Dietary Shifts of the Neolithic and Industrial Revolutions," *Nature Genetics* 45 (2013): 450–55.

17. Escaping the Laws of Nature

1. The idea apparently derives from the book of Exodus in which God gave Moses "two tables of testimony, tables of stone, written with the finger of God." Those words were to be "kept in the heart of man."

2. J. Lehrer, "A Physicist Solves the City," *New York Times*, December 19, 2010.

3. This required the physicist (or at least his students and assistants) to do something new: they had to comb census databases and other sources to understand the features of different blocks, neighborhoods, cities, and countries. They gathered data on density, walking speed, purchases, and nearly anything else they could find. If they did not consider clown density, for example, that is just because they couldn't find any way to compile the data.

4. L. M. A. Bettencourt et al., "Urban Scaling and Its Deviations: Revealing the Structure of Wealth, Innovation and Crime Across Cities," *PLoS One* 5 (2011): 1–9.

5. John Whitfield, *In the Beat of a Heart: Life, Energy, and the Unity of Nature* (Washington, DC: Joseph Henry Press, 2005).

6. Importantly, heart rate increases more slowly than body size because larger animals need progressively less active hearts, so the line relating the two variables is a curve unless one of the axes is logarithmic.

7. You might think, Well, why doesn't the body just make bigger arteries and veins and produce more blood rather than altering its heart rate as a function of size? Here is where one of West's great insights comes in. Because the number of capillaries is set by the size of an organism and their width is invariant, the total cross section of the vascular system is determined by body size. The body actually has no ability, evolutionarily, to change artery and vein size. The only way to pump more or less blood is by changing the heart rate.

8. An interesting aside here is the question of why mammals and birds need to stay warm in the first place. One camp suggests it allows them to be more active and get to the food, mates, and whatever else first. Another, though, suggests that warm-bloodedness evolved as a way to kill off fungal and other pathogens that cannot deal with body heat. Interestingly, the body temperature of the majority of mammals is just high enough to kill most fungi but not high enough to prove fatal to the mammals' own cells.

9. Recently, this model has been fine-tuned. The wear and tear on our cells comes largely from the production of free radicals of oxygen. Those free radicals bang around in the cells and damage things, but they are produced during metabolism and so are inescapable. But just how bad their effect is seems to be mediated in part by the behavior of mitochondria. In some cells and animals, mitochondria spend most of their time producing heat rather than energy. When they do so, they generate fewer free radicals. As a consequence, they might also generate less wear and tear, and so it might be the case that animals whose mitochondria spend more time producing heat and less time producing energy live a little longer than would be predicted based on their heart rates, though this remains to be seen.

10. H. J. Levine, "Rest Heart Rate and Life Expectancy," *Journal of the American College of Cardiology* 30 (1997): 1104–6.

11. Biologists used to think the same until, in 1970, the bear biologist Lynn Rogers began to study hibernating bears with questions of heart rate in mind. Rogers does dangerous things with bears. He follows his favorite ones through the forest, walking behind them as though on a walk with his favorite dog. They are used to him, mostly. Lynn spent more time with bears than anyone, aside from other bears. Who better to study how they hibernate. How do you study hibernating bears? Rogers crawled into a bear cave with an anal heart-rate monitor and probed several bears without event. He even put his head up to the hairy, stinking body of one hibernat-

ing bear. She did not stir; he could barely hear her heart. A few bears lifted their heads. One growled. But mostly they stayed still, inert. Then, on March 27, 1970, Rogers fell on a six-year-old female bear. Hibernating animals slept, it was thought at the time, so deeply as to seem harmless, but the baby alongside this six-year-old woke up. Still, the mother did not stir. Rogers couldn't resist poking her. Could she be woken? He poked her some more, and suddenly her heart went from quiet to racing, 175 beats per minute—faster than had been recorded even from very active bears. The mother bear was not happy. That is when Lynn discovered two more things: that bears wake rapidly from their winter slumber and that, for a short distance, he could outrun a pair of sleepy but angry bears.

12. S. Telles et al., "An Evaluation of the Ability to Voluntarily Reduce the Heart Rate After a Month of Yoga Practice," *Integrative Physiological and Behavioral Science* 39 (2004): 119–25.

13. Interestingly, the influence of yoga on heart rate, although it seems pronounced, has been little studied. This paper appears to be the last one published on the topic, and it came out in 2004.

14. Evolutionarily, this is an accurate impression. They belong in the family of the squirrels, Sciuridae.

15. Yes, miniature elephants actually did exist.

16. B. W. Johansson, "The Hibernator Heart—Nature's Model of Resistance to Ventricular Fibrillation," *Arctic Medical Research* 50 (1991): 58–62.

17. I. Oransky, "Wilfred Gordon Bigelow," *Lancet* 365 (2005): 1616. In the paper that resulted from the talk, Bigelow was forced to note, "Since this article was written, Drs. Charles Bailey of Philadelphia and F. J. Lewis of Minneapolis have reported the successful use of this technic of cooling upon 2 patients." Bigelow's insight had, within just months, resulted in a successful heart surgery.

18. W. G. Bigelow and J. E. McBirnie, "Further Experiences with Hypothermia for Intracardiac Surgery in Monkeys," *Annals of Surgery* 37 (1965): 361–65.

19. G. W. Miller, *King of Hearts: The True Story of the Maverick Who Pioneered Open Heart Surgery* (New York: Times Books, 2000).

20. Caffeine is similar enough to adenosine to bind its receptor and, thus, block the hibernation molecules from having their effect. In drinking coffee, you are convincing your body to ignore the signs that night or the long winter has arrived.

Postscript: The Future Science of the Heart

1. *Failure* is perhaps too strong a word because every medical failure spawns some innovation. The effort to transplant hearts, for example, pushed heart surgery forward, perhaps faster than it might have gone otherwise.

2. A. Abbott, "Doubts Over Heart Stem-Cell Therapy," *Nature* 509 (2014): 15–16.

3. J. B. Andersen et al., "Physiology: Postprandial Cardiac Hypertrophy in Pythons," *Nature* 434 (2005): 37–38; S. M. Secor and J. Diamond, "A Vertebrate Model of Extreme Physiological Regulation," *Nature* 395 (1995): 659–62.

4. C. A. Riquelme et al., "Fatty Acids Identified in the Burmese Python Promote Beneficial Cardiac Growth," *Science* 334 (2011): 528–31.

BIBLIOGRAPHY

Abbott, Elizabeth. *An Inner Grace: The Life Story of Dr. Maude Abbott and the Advent of Heart Surgery.* Amazon Digital Services, 2010.

Ashcroft, F. *The Spark of Life: Electricity in the Human Body.* New York: W. W. Norton, 2012.

Beattie, Andrew, et al. *Wild Solutions: How Biodiversity Is Money in the Bank.* New Haven: Yale University Press, 2001.

Bigelow, Wilfred Gordon. *Cold Hearts: The Story of Hypothermia and the Pacemaker in Heart Surgery.* Toronto: McClelland and Stewart, 1984.

Boring, Mel, et al. *Guinea Pig Scientists: Bold Self-Experimenters in Science and Medicine.* New York: Henry Holt and Co., 2005.

Capra, Fritjof. *Learning from Leonardo: Decoding the Notebooks of a Genius.* San Francisco: Berrett-Koehler, 2013.

———. *The Science of Leonardo.* New York: Random House, 2007.

Cobb, W. M. *The First Negro Medical Society.* Washington, DC: Associated Publishers, 1939.

Cohen, J. *Almost Chimpanzee: Searching for What Makes Us Human.* New York: Henry Holt and Company, 2010.

Cohn, S. *It Happened in Chicago.* Guilford, CT: Globe Pequot Press, 2009.

Cooney, K. *The Woman Who Would Be King.* New York: Crown Publishing, 2014.

Cooper, D. K. C. *Open Heart: The Radical Surgeons Who Revolutionized Medicine.* New York: Kaplan Publishing, 2010.

Evans, Arthur V., et al. *An Inordinate Fondness for Beetles.* Oakland: University of California Press, 2000.

Forssmann, F. *Experiments on Myself: Memoirs of a Surgeon in Germany.* New York: St. Martin's Press, 1974.

Greatbatch, W. *The Making of the Pacemaker: Celebrating a Lifesaving Invention.* New York: Prometheus Books, 2000.

Halperin, J. L., and R. Levine. *Bypass*. New York: Times Books, 1985.

Hamilton, D. *The Monkey Gland Affair*. London: Chatto and Windus, 1986.

Hankinson, R. J. (ed.). *The Cambridge Companion to Galen*. Cambridge: Cambridge University Press, 2009.

Harvey, William. *On the Motion of the Heart and Blood in Animals*. Translated by Robert Willis. New York: Prometheus Books, 1993.

Hollingham, R. *Blood and Guts: A History of Surgery*. New York: St. Martin's Press, 2009.

Hurt, R. *The History of Cardiothoracic Surgery: From Early Times*. New York: CRC Press, 1996.

Jones, David S. *Broken Hearts: The Tangled History of Cardiac Care*. Baltimore: Johns Hopkins University Press, 2014.

Keys, Ancel. *How to Eat Well and Stay Well the Mediterranean Way*. New York: Doubleday, 1975.

Keys, Ancel, and Margaret Keys. *Eat Well and Stay Well*. New York: Doubleday, 1963.

Keynes, Geoffrey. *The Life of William Harvey*. Oxford: Oxford University Press, 1966.

Kirk, J. *Machines in Our Hearts: The Cardiac Pacemaker, the Implantable Defibrillator, and American Health Care*. Baltimore: Johns Hopkins University Press, 2001.

Kuss, R., and P. Bourget. *Une histoire illustrée de la greffe d'organes: La grande aventure du siècle*. Librairie Sandoz et Fischbacher (openlibrary.org), 1992.

Lax, Eric. *The Mold in Dr. Florey's Coat: The Story of the Penicillin Miracle*. New York: Henry Holt and Co., 2004.

Le Fanu, James. *The Rise and Fall of Modern Medicine*. New York: Basic Books, 2012.

Lester, Toby. *Da Vinci's Ghost: Genius, Obsession, and How Leonardo Created the World in His Own Image*. New York: Simon and Schuster, 2012.

Li, Jie Jack. *Triumph of the Heart: The Story of Statins*. New York: Oxford University Press, 2009.

Malloch, Archibald. *William Harvey*. Whitefish, MT: Kessinger, 2010.

Malloy, L. "Helen Brooke Taussig (1898–1986)," in J. Bart, ed., *Women Succeeding in the Sciences: Theories and Practices Across Disciplines*. West Lafayette, IN: Purdue University Press, 1999.

Mattern Susan P. *The Prince of Medicine: Galen in the Roman Empire*. Oxford: Oxford University Press, 2013.

McRae, Donald. *Every Second Counts: The Race to Transplant the First Human Heart*. New York: Putnam, 2006.

Meriwether, Louise. *The Heart Man: Dr. Daniel Hale Williams*. New Jersey: Prentice-Hall, 1972.

Miller, G. Wayne, *King of Hearts: The True Story of the Maverick Who Pioneered Open Heart Surgery*. New York: Broadway Books, 2000.

Monagan, David. *Journey into the Heart: A Tale of Pioneering Doctors and Their Race to Transform Cardiovascular Medicine*. New York: Gotham Books, 2007.

Money, Nicholas P. *Mr. Bloomfield's Orchard: The Mysterious World of Mushrooms, Molds, and Mycologists*. Oxford: Oxford University Press, 2004.

Moore, F. D. *A Miracle and a Privilege: Recounting a Half Century of Surgical Advance*. Washington, DC: National Academy Press, 1995.

Morris, Charles R. *The Surgeons: Life and Death in a Top Heart Center*. New York: W. W. Norton, 2008.

Nunn, John F. *Ancient Egyptian Medicine*. Norman: University of Oklahoma Press, 2002.

Orrin, H. C. *The X-ray Atlas of the Systemic Arteries of the Body*. New York: Baillière, 1920.

Robinson, D. B. *The Miracle Finders: The Stories Behind the Most Important Breakthroughs of Modern Medicine*. New York: David McKay, 1976.

Sawday, Jonathan. *The Body Emblazoned: Dissection and the Human Body in Renaissance Culture*. New York: Routledge, 1996.

Schmid-Hempel, Paul. *Parasites in Social Insects*. Princeton: Princeton University Press, 1998.

Seaborg, Glenn T. *Adventures in the Atomic Age: From Watts to Washington*. New York: Farrar, Straus and Giroux, 2001.

Sedmera, David, and Tobias Wang (eds.). *Ontogeny and Phylogeny of the Vertebrate Heart*. New York: Springer, 2012.

Shelley, Mary. *Frankenstein*. Edited by Maurice Hindle. New York: Penguin, 2005.

Shubin, Neil. *Your Inner Fish*. New York: Vintage, 2009.

Slights, William W. E. *The Heart in the Age of Shakespeare*. New York: Cambridge University Press, 2011.

Taussig, Helen B. *Congenital Malformations of the Heart*. Cambridge, MA: Harvard University Press, 1960.

Tilney, Nicholas L., *Invasion of the Body*. Cambridge, MA: Harvard University Press, 2011.

Tucker, T. *The Great Starvation Experiment*. New York: Free Press, 2006.

Varki, Ajit, and Danny Brower. *Denial: Self-Deception, False Beliefs, and the Origins of the Human Mind*. New York: Twelve, 2013.

Varki, Ajit, et al. (eds.). *Essentials of Glycobiology*, 2d ed. New York: Cold Spring Harbor Laboratory Press, 2009.

Vogel, Steven. *Prime Mover: A Natural History of Muscle*. New York: W. W. Norton, 2003.

Weisse, Allen B. *Heart to Heart—The Twentieth-Century Battle Against Cardiac Disease: An Oral History*. New Brunswick, NJ: Rutgers University Press, 2002.

Wells, Francis. *The Heart of Leonardo*. New York: Springer, 2013.

———. "The Renaissance Heart," in J. Peto, ed., *The Heart*. New Haven: Yale University Press, 2007.

Whitfield, John. *In the Beat of a Heart: Life, Energy, and the Unity of Nature*. Washington DC: Joseph Henry Press, 2005.

INDEX

Note: *Italic* page numbers refer to illustrations.

ABOUT THE AUTHOR

Robb Dunn is an associate professor in Ecology and Evolution in the Department of Biological Sciences at North Carolina State University. He's the author of *The Wild Life of Our Bodies* and *Every Living Thing*, and his magazine work is published widely, including in *National Geographic*, *Natural History*, *New Scientist*, *Scientific American*, and *Smithsonian*. He has a PhD from the University of Connecticut and was a Fulbright Fellow. He lives in Raleigh, North Carolina.